STATISTICAL CALCULATION FOR BEGINNERS

STATISTICAL CALCULATION FOR BEGINNERS

BY

E. G. CHAMBERS, M.A.
Assistant Director of Research in Industrial Psychology, Cambridge; Fellow of the Royal Statistical Society

CAMBRIDGE
AT THE UNIVERSITY PRESS
1958

CAMBRIDGE UNIVERSITY PRESS
Cambridge, New York, Melbourne, Madrid, Cape Town, Singapore, São Paulo, Delhi

Cambridge University Press
The Edinburgh Building, Cambridge CB2 8RU, UK

Published in the United States of America by Cambridge University Press, New York

www.cambridge.org
Information on this title: www.cambridge.org/9780521116206

First edition 1940
Reprinted 1943, 1945, 1946, 1948
Second edition 1952
Reprinted 1955, 1958
This digitally printed version 2009

A catalogue record for this publication is available from the British Library

ISBN 978-0-521-04613-8 hardback
ISBN 978-0-521-11620-6 paperback

Preface to Second Edition

In the ten years that have elapsed since this book was first published, certain additional statistical methods of particular value to research workers have been developed. These methods, presented as simply as possible, are included in this new edition. The whole of the original text has been revised and some small arithmetical errors corrected. I am grateful to those who have pointed out these mistakes.

The original numbering of the chapters and sections has been preserved as far as possible. The chief additional material is as follows. An account of the nature and use of the binomial distribution is given in Chapter IV. The calculation of t from arbitrary origins, of the variance ratio and of the standard error of the difference between proportions is included in Chapter V. In Chapter VII the Kendall coefficient of rank correlation is introduced and in Chapter IX a method of fitting linear and logarithmic curves to observational data is explained. Chapters X and XI are entirely new, the latter giving a short introduction to the method of analysis of variance. There are also some hitherto unpublished tables in the Appendices F–K. Exercises on the new material are provided, with answers. I am especially grateful to Mr J. W. Whitfield and Miss V. R. Cane for helpful comments and criticisms.

E. G. C.

Cambridge
June 1950

Preface to First Edition

The purpose of this book is to explain as simply as possible how to perform the calculations involved in the commoner statistical methods. It is not in any sense a treatise on the theory of statistics, only sufficient theory being given to enable the student to understand the use and application of the methods described.

No assumption of mathematical ability on the part of the reader is made; the calculations described involve the use of arithmetic only. A worked example of each method given is provided and abundant exercises with answers are supplied.

Whilst the book is chiefly addressed to students of the biological sciences, especially Psychology, the methods described are fundamental to statistical work and should, it is hoped, prove useful to anyone who has to make use of elementary statistical methods.

I should like to express my sincere gratitude to Dr J. O. Irwin, who very kindly read the whole of the manuscript and made many extremely helpful criticisms, and to Dr J. Wishart, who made some very valuable suggestions when the manuscript was approaching its final form. I wish to acknowledge also the kindness of Professor R. A. Fisher and his publishers, Messrs Oliver and Boyd, in allowing me to print extracts from various statistical tables given in his book *Statistical Methods for Research Workers*, and in the statistical tables by Fisher and Yates, also published by Oliver and Boyd. Full references to these two works are made in the text.

E. G. CHAMBERS

Cambridge
August 1940

Contents

Chapter I

INTRODUCTION

1.1. Description of certain statistical terms. The statistical methods described in this book are all concerned with the treatment of *variables*. By a variable is meant a quantity which assumes different values that may be measured in some appropriate unit. Height, weight, test scores, readings on a thermometer, etc., are examples of variables, or variates, as they are sometimes called. Variables are usually denoted by X or Y in this book. The number of times a particular value of a variable occurs in a set of observations is called the *frequency* of occurrence of that value, and a table showing the frequency of occurrence of all the values of a variable in a set of observations is named a *frequency distribution table*.

A series of observations may be represented by one value which is called an *average*, and the way in which the different values of the variable lie about this average is described as the *scatter* or *dispersion* of the observations. Measures of averages and scatter are descriptive statistics, since they yield in a condensed form a description of a whole series of observations. This is the first function of statistical method: the other chief function is the examination of various hypotheses which are made about observational data.

It is usually impossible to measure all the values of any variable, so that the data from a single experiment are only a *sample* drawn from the *total population* of possible observations. For example, if the variable is human height, then the total population of that variable would be the height of every man, woman and child ever on earth; it is manifestly impossible to measure all these values and in practice we have to content ourselves with measuring the heights of a sample of some convenient size. The distribution of the total population can usually be expressed in a mathematical form by using a small number of constants or *parameters*. Obviously we can never know the exact values of these parameters, since we cannot measure the whole population, but we can make estimates of them by measuring *random samples*, that is, samples drawn purely at random from the population. These estimates are known as *statistics*, and their accuracy as estimates depends on the size of the sample and the type of distribution of the variable.

In calculating some statistics it is essential to know the number of *degrees of freedom* available for the calculation. The conception of 'degrees of freedom' is not an easy one for the beginner, so that whenever the term is used in this book categorical rules for determining the number of degrees of freedom are supplied. Consideration of the following example may give some idea of the meaning of the term. Suppose 100 shillings are to be shared amongst 10 boys. We may give as many as we like (up to a total of 99) to each of 9 of the boys, but we are bound to give the tenth boy what is left over, i.e. we have only 9 degrees of freedom in sharing out the shillings. If we were told further that 60 shillings were to be shared amongst the oldest 5 boys and the remainder amongst the youngest 5, we should only have 8 degrees of freedom for doing this, since the fifth boy in each group would have to have what was left over —we should not be free to vary his share.

Two variables which are related together, so that a knowledge of the values of one variable indicates likely values of the other, are said to be *associated* or *correlated*. If the variables are unrelated, they are said to be *independent*.

Other terms, applicable to particular methods, will be described in their appropriate places.

1.ii. Notation. A certain amount of symbolism is essential in the description of statistical methods. Unfortunately there is a lack of agreement amongst different authors, which is apt to be confusing to the beginner. For this reason an attempt at a consistent method of notation is made in this book. Being based on first principles it is hoped that it will be readily understood by the learner and will enable him to follow the notation used in the standard books on statistical methods. Symbols in general established use are taken over unchanged. Here again there is confusion, since owing to the multiplicity of statistics the same symbol may have to stand for quite different quantities. Thus the symbol z is used for three different quantities and particular care is needed on the part of the student to avoid misconception in such cases.

Certain symbols have the same significance throughout the whole of this book. For example, N always stands for the total number of observations in a sample and S always signifies 'the sum of'. The other notation is made as unambiguous as possible. Certain Greek letters are used as symbols: a list of these with their pronunciation is given below:

β (beta)	π (pi)
γ (gamma)	ρ (rho)
δ (delta)	σ (sigma)
ζ (zeta)	τ (tau)
η (eta)	ϕ (phi)
μ (mu)	χ (ki)
ν (nu)	

Σ (capital sigma) is used in some places to indicate 'the sum of'. As a rule S is used for summation of sample values and Σ for derived quantities.

Care must be taken by the student to avoid confusing *suffixes* and *indices*. Suffixes are small numbers or letters written after a symbol at the foot, e.g. x_1, σ_x, etc.; these are merely descriptive and confine the use of the symbol to a particular purpose. Indices are small numbers written after and above symbols and have their usual algebraical significance; for example, x^2 (x squared) means x multiplied by x, y^3 (y cubed) means y multiplied by y multiplied by y, and so on.

The usual arithmetical symbols, $+$, $-$, $\times$ and $\div$, have their accustomed significance. There are three other symbols with which the non-mathematical student may not be familiar. Vertical lines drawn on each side of a quantity mean 'the positive numerical value of ', e.g. $|a-b|$ means 'the positive numerical value of the difference between a and b'. Using this notation, therefore, it does not matter whether we write $|a-b|$ or $|b-a|$. Secondly, there is the factorial sign, '!'. This latter is best explained by examples, e.g. 4! stands for $4 \times 3 \times 2 \times 1$, 6! for $6 \times 5 \times 4 \times 3 \times 2 \times 1$, and so on. Thirdly, $\binom{n}{p}$ means the number of combinations of n things taken p at a time. Expanded algebraically, $\binom{n}{p} = \frac{n!}{p!(n-p)!}$.

Since a good deal of arithmetical work is involved in certain of the statistical methods described in the following chapters, it is an advantage for the student to be familiar with the use of logarithms (unless he has a calculating machine available).

1.iii. References. Reference is made in the following pages to two invaluable books on statistics and to certain books of statistical tables. The references are made numerically to the following works:

(1) *An Introduction to the Theory of Statistics.* G. Udny Yule and M. G. Kendall. 11th edition. Charles Griffin and Co. Ltd. 1937.
(2) *Statistical Methods for Research Workers.* R. A. Fisher. 9th edition. Oliver and Boyd. 1944.
(3) *Tables for Statisticians and Biometricians*, Part I. Edited by Karl Pearson. Biometrika Office, University College, London.
(4) *Barlow's Tables of Squares, Cubes, Square-roots, Cube-roots and Reciprocals of all Integral Numbers up to* 12,500. E. and F. N. Spon. 4th edition. 1941.
(5) *Statistical Tables for Biological Agricultural and Medical Research.* Fisher and Yates. Oliver and Boyd. 2nd edition. 1942.
(6) *The Advanced Theory of Statistics.* M. G. Kendall. Charles Griffin and Co. Ltd. Vol. I, 1945; vol. II, 1946.

Since no attempt is made in this book to prove or justify the various methods and formulae used, the student wishing to go into such matters is referred to the first two and the last of the foregoing works.

1.iv. Use and abuse of statistical methods. The student who works conscientiously through the following chapters should learn how to make use of the commoner methods of statistics. He should never forget, however, that statistical methods are merely tools for a research worker. They enable him to describe, relate and assess the value of his observations. They cannot make amends for incorrect observation nor can they of themselves provide a single fact of psychology, biology or any other subject of research. Statistical methods are to the research worker what tools are to a carpenter. The latter has first to learn how to use his tools and he may then by employing them reveal the useful and beautiful purposes to which his material may be put. But the tools themselves must be used for their correct functions. The craftsman will not, for instance, use a mallet and chisel or a fretsaw to plane a plank of wood, nor will he use a hammer to drive in a screw. In the same way statistical methods must only be used by the research worker for the purposes for which they have been devised.

Further, a carpenter's tools cannot tell him directly anything about the materials he is using. They cannot by themselves distinguish between mahogany and deal nor prove that oak is more durable than white wood. No carpenter's tools have ever yet made a piece of wood; similarly no statistical method has ever yet produced a biological fact.

The student is advised, therefore, to try to acquire an understanding of the specific purpose of each statistical method he learns to use, to appreciate the scope of and the assumptions underlying the use of each formula, and to realise that the outcome of each calculation is a statistical statement which has to be interpreted in terms of the particular branch of science from which the data for examination are drawn.

Chapter II

AVERAGES

2.i. The arithmetic mean. The best known and most useful form of average is the *arithmetic mean*, usually referred to as the 'mean' or the 'average'. It is easily calculated by adding together all the observations to be averaged and dividing the sum or total by the number of observations.

Example 1. Find the mean of the following observations: 22, 24, 20, 23, 21, 19, 23, 22, 20, 22, 20, 22, 23, 25, 21, 21, 22, 24, 23, 22, 23, 21, 22, 21, 23.

Add together all the observations.

The sum $= 549$,

The number of observations $= 25$,

$$\text{The arithmetic mean} = \frac{\text{sum of observations}}{\text{no. of observations}} = \frac{549}{25} = 21{\cdot}96.$$

This procedure may be generalised to cover all cases. If X is a variable which has different values X_1, X_2, X_3, etc., then the arithmetic mean of a number N of such values is the sum of the various values of X, which we denote by $S(X)$, divided by N, the number of them. In general, therefore,

$$m_x = \bar{X} = \frac{S(X)}{N}. \tag{1}$$

Here m_x and $\bar{X}$ (called X-bar) are different ways of denoting 'the mean of X'.

2.ii. If N is large and no adding machine is available, the process of addition may be very laborious. It may, however, be made easier by the construction of a *frequency distribution table*. This is a table showing how often each value of the variable occurs in the sample under consideration. In Example 1, the values taken by X all lie between 19 and 25 inclusive. If we count how many times each different value of X occurs and write the totals in tabular form, we obtain the frequency distribution table given below.

TABLE I

X	f
19	1
20	3
21	5
22	7
23	6
24	2
25	1
	25

In this table the first column, headed X, shows the different values assumed by the variable X in the sample, and the second column, headed f, gives the number of times, or frequency, of occurrence of each. The total of the f column is, of course, the total number of observations we are averaging, i.e. $S(f) = N$. The next step is to write down a third column, headed fX, which is produced by multiplying together the corresponding pairs of numbers in the X and f columns. We then sum the fX column, giving us $\Sigma(fX)$, and the arithmetic mean is then obtained by dividing this sum by N as before; i.e.

$$m_x = \bar{X} = \frac{\Sigma(fX)}{N}. \tag{2}$$

Example 2. Calculate the mean of the observations in Example 1 by constructing a frequency of distribution table.

X	f	fX
19	1	19
20	3	60
21	5	105
22	7	154
23	6	138
24	2	48
25	1	25
	25	549

$$\Sigma(fX) = 549,$$
$$N = 25,$$
$$\bar{X} = \frac{549}{25} = 21{\cdot}96.$$

It will be noted that this result is identical with that obtained in Example 1.

2.iii. The method of Section 2.ii is useful when the range of the X values is small, but if there are many different values of X the method again becomes laborious. Suppose the observations in Example 1 were the lengths of 25 sticks measured in centimetres, each one being measured to the nearest centimetre. Now let us suppose we had a large number of such sticks and measured them in millimetres. The range of the measurements might now be from 190 to 252 mm., so that if we constructed a frequency distribution table of these we should have 63 different values of X to tabulate. This would

TABLE II

X	f
190–193	2
194–197	4
198–201	7
202–205	12
206–209	19
210–213	24
214–217	27
218–221	35
222–225	26
226–229	21
230–233	18
234–237	13
238–241	6
242–245	5
246–249	2
250–253	1
	222

be tedious, and the calculation may be shortened, with some small sacrifice of accuracy, by subdividing the range of the X's into a convenient number of groups. In practice, a number of groups between 12 and 20 should be chosen, and the best unit for grouping may be found by dividing the range first by 12 and then by 20, and taking a convenient unit in between these two quotients. For instance, if the range is 63, then the results of dividing 63 first by 12 and then by 20 are 5·25 and 3·15. Hence a convenient working unit for grouping would be 4 in this case. This means that we should group the values of

X together in 4's, so that the first group would comprise 190, 191, 192 and 193, the second 194, 195, 196 and 197, and so on, the last group being 250, 251, 252 and 253. We could now construct a frequency distribution table of these groups. Such a table might be, for example, as Table II. This gives the frequency distribution of the lengths of 222 sticks measured in 4 mm. groups.

Now the method of calculating the mean length of the sticks from such a table depends on the assumption that the average length of the sticks in each group is equal to the mean value of X for that group. For instance, there are 12 sticks in the 202–205 group and we shall assume that the average length of those 12 sticks is equal to the average of 202, 203, 204 and 205, i.e. 203·5. This is, of course, an assumption, but the larger the number of readings in each group the nearer it becomes to being true.

Care should be taken in arranging the grouping that this assumption should be as nearly as possible true for the end groups, leaving the middle ones, which usually have more readings in them, to look after themselves. For instance, if the two readings in the 190–193 group were both 190, then their average would be 190 instead of 191·5, which is assumed by the grouping. In such a case it would have been better to have started the grouping from 189, which would assume an average for the group of 190·5.

In order to calculate the mean of the observations in a grouped frequency distribution table, such as Table II, we take an *arbitrary origin*, or starting point, and then calculate the discrepancy between this point and the true mean. Let us take our arbitrary origin near the middle of the range, since this simplifies the arithmetic. It is convenient to have it at the centre of a group so we will choose the 218–221 group, so that our arbitrary origin will be 219·5, the centre of the group. This group we number 0. The next group in the table, 222–225, has an average of 223·5, which is one group unit, or working unit, above the arbitrary origin, and we therefore number this group 1. In a similar manner the 226–229 group is numbered 2, and so on.

The 214–217 group averages 215·5, which is one working unit less than the arbitrary origin, so this group is numbered −1. Similarly the 210–213 group becomes −2, and so on.

Example 3. Calculate the mean of the data in Table II.

X	f	x	fx
190–193	2	−7	−14
194–197	4	−6	−24
198–201	7	−5	−35
202–205	12	−4	−48
206–209	19	−3	−57
210–213	24	−2	−48
214–217	27	−1	−27
218–221	35	0	−253*
222–225	26	1	26
226–229	21	2	42
230–233	18	3	54
234–237	13	4	52
238–241	6	5	30
242–245	5	6	30
246–249	2	7	14
250–253	1	8	8
	222		256
			−253
			3

$$\Sigma(fx) = 3,$$
$$N = 222,$$
$$D = \frac{3}{222} = 0{\cdot}0135,$$
$$w = 4,$$
$$m_a = 219{\cdot}5,$$
$$m_x = 219{\cdot}5 + 0{\cdot}0135 \times 4$$
$$= 219{\cdot}5 + 0{\cdot}054$$
$$= 219{\cdot}554.$$

* Since there will be no entry in the fx column corresponding to $x = 0$, this is a convenient place to add the negative entries in the fx column.

We can now replace the X column by another column, which we will head 'x', which indicates the number of working units that each X-group lies away from the arbitrary origin. The X column is now neglected and a fourth column, headed fx, is written down. This is obtained by multiplying corresponding entries in the f and x columns. By adding this column we get $\Sigma(fx)$, and the discrepancy, D, between the true mean and the arbitrary origin, *in working units*, is given by

$$D = \frac{\Sigma(fx)}{N}. \tag{3}$$

This quantity D tells us how many working units the true

mean m_x lies away from the arbitrary origin, which we will call m_a. Hence the true mean is given by the formula

$$m_x = \bar{x} = m_a + Dw, \tag{4}$$

where w is the size of the working unit. If D turns out to be negative, then Dw will have to be subtracted from m_a. If it should happen that the arbitrary origin is the true mean, as might happen with a perfectly symmetrical distribution, then D will, of course, be zero.

The whole process of calculating the mean by this method is shown in Example 3.

We conclude therefore that the mean length of the sticks was 220 mm., to the nearest millimetre.

2.iv. The median. Another form of average which is sometimes used for convenience is the *median*. This, as its name implies, is the middle observation and it is easily found in ungrouped data by ranking the sample of observations in order of their size and finding the central observations, if N is odd, or the mean of the two observations in the middle, if N is even. If N is odd, the median will be the $(N+1)/2$th observation. If N is even, it will be the mean of the $N/2$th and the $N/2+1$th observations. For example, the median of 2, 4, 6, 8, 10, 12 is the mean of the 3rd and 4th readings, i.e.

$$(6+8)/2 = 7.$$

The median is representative of a set of observations in the sense that there are exactly as many observations greater than it as there are less. If the distribution is perfectly symmetrical, the median is equal to the mean.

Example 4. Find the median of the observations in Example 1.

Ranking the observations in order of size we get: 19, 20, 20, 20, 21, 21, 21, 21, 21, 22, 22, 22, 22, 22, 22, 22, 23, 23, 23, 23, 23, 23, 24, 24, 25.

Since there are 25 observations, the median will be the 13th, i.e. 22.

2.v. Finding the median for grouped data involves a little approximation, as was the case in finding the mean of grouped data. We assume that the values in each group are evenly distributed through the group and the approximate value of the median may be found by linear interpolation. Suppose, for example, we wished to find the median of the data in Table II. We have to find the value of the observation below which half the observations lie. N in this case is 222, so that $\frac{1}{2}N$ is 111. The median therefore will be the position of the mean of the 111th and 112th observations. By adding together the first seven entries in the f column we find that 95 of the observations lie below 217·5,* and there are 35 observations in the 218–221 group. Obviously, therefore, the median lies somewhere in this group. The position may be found by simple proportion. $111-95 = 16$. In the 218–221 group 16 observations will therefore be less and 19 greater than the median, assuming that all the values in this group are evenly distributed. The group contains 4 units, so that the position of the median is given by adding $16/35 \times 4$ to 217·5, the limit of the previous group. Hence the median of the data in Table II is

$$217{\cdot}5+\frac{16}{35}\times 4 = 217{\cdot}5+1{\cdot}83 = 219{\cdot}33.$$

This may be checked by working from the other end of the table. We find that 92 observations lie *above* 221·5, so that the median is

$$221{\cdot}5-\frac{111-92}{35}\times 4 = 221{\cdot}5-2{\cdot}17 = 219{\cdot}33.$$

2.vi. The mode. A form of average which is occasionally used is the *mode*. This is the most frequently occurring, or most fashionable, observation. For instance, in the data of Example 1, the mode would be 22, since this observation occurs more often than any other. However, the mode cannot usually be found as easily as this for small samples, since errors of sampling may result in the frequent occurrence of some observation remote from the true mode.

* Note that since the measurements in the X column are made to the nearest millimetre, any observation up to but not equal to 217·5 will count as 217 or under, so that the real upper limit of the group is 217·5. Similarly the real lower limit of the 222–225 group is 221·5.

If a graph of the frequency distribution of any variable for the total population were drawn, we should have a smooth curve (see Fig. 2, p. 25 for an example). In such a curve the position of the mode would be given by the highest point, since there would be the maximum frequency at that point. Such a curve might be symmetrical or it might be asymmetrical, or *skew*. It has been found that in curves of moderate skewness the position of the mode is given approximately by the relation

$$\text{Mode} = \text{mean} - 3\,(\text{mean} - \text{median}).$$

It is better, therefore, to make use of this equation for finding the mode of a sample than to rely on picking out the most frequently occurring observation. (We are considering here only distributions which give a hump-backed curve and have therefore only one mode. Certain distributions may have two or more modes, but the elementary student need not concern himself with these.)

Example 5. Find the mode of the data in Table II.

We have found previously that the mean of the data is 219·55 and the median is 219·33. Substituting these values in the equation

$$\text{Mode} = \text{mean} - 3\,(\text{mean} - \text{median}),$$

we have

$$\begin{aligned}\text{Mode} &= 219{\cdot}55 - 3\,(219{\cdot}55 - 219{\cdot}33)\\ &= 219{\cdot}55 - 0{\cdot}66\\ &= 218{\cdot}89.\end{aligned}$$

2.vii. It may be seen from the above equation that if the mean and median are equal the mode is also equal to the mean. Thus in the case of a variable which is distributed in a perfectly symmetrical manner, the mean, the median and the mode are all equal.

Note on the construction of a frequency distribution table

If the observations to be tabulated are on cards, the frequency distribution table is easily formed by sorting the cards into their appropriate groups and counting the number of cards in each group. When, however, the data are not on cards, a frequency table may be made by the use of a *spot diagram*. This is made by tabulating the X column and then going through the observations one at a time and putting a spot against the appropriate X-group for each observation. The data in Table II would appear as under in a spot diagram.

Spot diagram of the data in Table II

X	f
190–193	2
194–197	4
198–201	7
202–205	12
206–209	19
210–213	24
214–217	27
218–221	35
222–225	26
226–229	21
230–233	18
234–237	13
238–241	6
242–245	5
246–249	2
250–253	1
	222

It helps in counting the spots for the entries in the f column if they are put down in groups of five, as above.

EXERCISES ON CHAPTER II

1. Find the arithmetic mean of the scores in each of the tests in Appendix E from A to G inclusive for

(*a*) subjects 1–25,
(*b*) subjects 26–50,
(*c*) subjects 51–75,
(*d*) subjects 76–100.

Use formula (1).

2. Find the arithmetic mean of the scores in test F for

(*a*) subjects 1–50,
(*b*) subjects 51–100.

Use formula (2).

3. Construct grouped frequency distribution tables for the scores in each of the tests A to G inclusive for the whole 100 subjects. Use the following group units and starting points.

Test	Group unit	First group
A	4	14–17
B	15	122–136
C	4	0–3
D	4	4–7
E	6	78–83
F	2	8, 9
G	3	17–19

(*a*) Thence, using formulae (3) and (4), calculate the mean scores in each of the tests A to G inclusive for the whole 100 subjects.

(*b*) Compare the means obtained with those given by adding the means of the four groups of 25 subjects found in Exercise 1 and dividing the totals by 4. These latter means are accurate and will indicate the discrepancies introduced by grouping the data.

4. Find the median of the scores in each of the tests from A to G inclusive for

(*a*) subjects 1–50,
(*b*) subjects 51–100.

5. From the answers to Exercise 1 compute the means of the scores of each test from A to G inclusive for

(*a*) subjects 1–50,
(*b*) subjects 51–100.

Then, making use of the answers to Exercise 4, estimate the mode of the scores in each test for

(*a*) subjects 1–50,
(*b*) subjects 51–100.

Use the relationship 'Mode = mean − 3 (mean − median)'.

Chapter III

SCATTER OR DISPERSION

3.i. Whilst an average to some extent represents the whole series of observations of which it is the mean, yet it does not by itself convey sufficient information about those observations. As a general rule it is also necessary to know how the observations are scattered around their average. Obviously the average of a given number of observations which all lie closely about the mean is more reliable as a representative statistic than one which is in the middle of a widely dispersed series of readings. It is important, therefore, to give an indication of the amount of scatter of the observations averaged.

3.ii. The range. The simplest measure of scatter or dispersion is the *range* of the observations, i.e. the distance between the largest and smallest observations. The range, however, is not a good measure of dispersion. It is based on two observations only, instead of on all the available information, and those two observations are liable to vary considerably in different samples, since the range depends essentially on the size of the sample.

Some writers give the range which includes all the observations except the 10 % smallest and the 10 % largest in an attempt to allow for the variability of the extremes of a sample, but this practice has little to recommend it.

3.iii. The inter-quartile range. A measure of dispersion which is not quite so subject to fluctuations of sampling is the *inter-quartile range*. This is obtained as follows. First rank all the observations in order of size and find the median (see section 2.iv). This divides the sample into two equal portions. Then find the median of each half: that of the lower half is called the *lower quartile*, that of the upper half the *upper quartile*. These two quartiles with the median divide the whole range of observations into four equal inter-quartile groups, and the distance between the lower and upper quartiles is called the inter-quartile range. It is usual to halve this and quote the 'semi-inter-quartile range'.

This measure of dispersion makes use of slightly more information than does the total range, but it does not lend itself to further mathematical treatment and is not a very valuable measure.

3.iv. The mean deviation. A measure of scatter which makes use of all the observations is obtained by writing down the difference between each separate observation and the average, adding together all these differences without regard to their signs, and dividing the total by the number of observations. This is called the *mean deviation* or *mean variation* and is best calculated from the median, as it is then a minimum.

Example 6. Calculate the mean deviation of the data in Example 1.

The median of the observations is 22, as was found in Example 4. The differences between each reading and the median, without regard to sign, are: 0, 2, 2, 1, 1, 3, 1, 0, 2, 0, 2, 0, 1, 3, 1, 1, 0, 2, 1, 0, 1, 1, 0, 1, 1.

$$\begin{aligned} &\text{The sum of these differences} &&= 27, \\ &N &&= 25, \\ &\text{The mean deviation} &&= \frac{27}{25} = 1{\cdot}08. \end{aligned}$$

3.v. The standard deviation. By far the best and most useful measure of scatter is the *standard deviation*. In words, this is the square root of the mean of the squares of the deviations of the observations from their arithmetic mean. In symbols, if $(X-\bar{X})$ represents the deviation of an individual reading from the mean and $S(X-\bar{X})^2$ the sum of the squares of all such deviations, then the standard deviation, σ, is given by the formula

$$\sigma = \sqrt{\frac{S(X-\bar{X})^2}{N}}. \tag{5}$$

The square of the standard deviation is called the *variance* or second moment, the latter usually being denoted by μ_2.

Hence the variance

$$= \sigma^2 = \mu_2 = \frac{S(X-\bar{X})^2}{N}. \qquad (5\,A)$$

The method of calculating the standard deviation depends on the type of data and the number of observations, and the following sections show how to calculate it in different cases.

3.vi. When N, the number of observations, is small and the mean of the observations is a whole number, the standard deviation is easily calculated by subtracting the mean from each observation, squaring each of these differences, adding the squares, dividing the sum of the squares by N and finally taking the square root of the quotient.*

Example 7. Find the standard deviation of the first 11 natural numbers.

X	$X-\bar{X}$	$(X-\bar{X})^2$
1	−5	25
2	−4	16
3	−3	9
4	−2	4
5	−1	1
6	0	0
7	1	1
8	2	4
9	3	9
10	4	16
11	5	25
66		110

$$S(X) = 66,$$
$$N = 11,$$
$$\bar{X} = \frac{66}{11} = 6,$$
$$S(X-\bar{X})^2 = 110.$$
$$\therefore \quad \sigma = \sqrt{\frac{110}{11}}$$
$$= \sqrt{10} = 3{\cdot}16.$$

3.vii. It rarely happens in practice, however, that the mean of a set of observations is a whole number or that N is as small as 11. If the mean is fractional, it would be laborious to square all the fractional differences between the readings and their mean. It is possible, however, to avoid this. If the different values of the variable are X_1, X_2, X_3, etc., then their

* When the variance in the whole population is being estimated from a sample, the sum of squares is divided by $N-1$ (see 5.ii, p. 37).

differences from the mean are $(X_1-\bar{X})$, $(X_2-\bar{X})$, $(X_3-\bar{X})$, etc. Squaring these we get

$$(X_1-\bar{X})^2 = X_1^2-2X_1\bar{X}+\bar{X}^2,$$
$$(X_2-\bar{X})^2 = X_2^2-2X_2\bar{X}+\bar{X}^2,$$
$$(X_3-\bar{X})^2 = X_3^2-2X_3\bar{X}+\bar{X}^2, \text{ etc.}$$

Summing these we have

$$S(X-\bar{X})^2 = S(X^2)-2\bar{X}.S(X)+N\bar{X}^2.$$

Divide both sides by N. Then

$$\frac{S(X-\bar{X})^2}{N} = \frac{S(X^2)}{N} - 2\bar{X}\frac{S(X)}{N} + \bar{X}^2.$$

But $\dfrac{S(X)}{N} = \bar{X}$, by formula (1); hence

$$\frac{S(X-\bar{X})^2}{N} = \frac{S(X^2)}{N} - 2\bar{X}^2+\bar{X}^2 = \frac{S(X^2)}{N} - \bar{X}^2. \qquad (5\text{ B})$$

In words, the variance is the mean of the squares of the observations less the square of their mean. Hence the standard deviation of the numbers in Example 7 could have been calculated as under:

X	X^2
1	1
2	4
3	9
4	16
5	25
6	36
7	49
8	64
9	81
10	100
11	121
66	506

$$S(X) = 66,$$
$$N = 11,$$
$$\bar{X} = 6.$$
$$\text{Variance} = \frac{506}{11} - 6^2$$
$$= 46-36$$
$$= 10.$$
$$\text{Hence} \quad \sigma = \sqrt{10} = 3{\cdot}16.$$

The method is more usefully extended to a frequency distribution table, for which we have the following formula:

$$\sigma^2 = \frac{\Sigma(fX^2)}{N} - \bar{X}^2. \qquad (6)$$

This may also be written

$$\sigma^2 = \frac{\Sigma(fX^2)}{N} - \left\{\frac{\Sigma(fX)}{N}\right\}^2. \tag{6 A}$$

The use of this formula is illustrated in the following example.

Example 8. Find the standard deviation of the data in the following frequency distribution table:

X	f
1	2
2	6
3	12
4	7
5	2
6	1

Construct two further columns, as under:

X	f	fX	fX^2
1	2	2	2
2	6	12	24
3	12	36	108
4	7	28	112
5	2	10	50
6	1	6	36
	30	94	332

$N = S(f) = 30,$

$\Sigma(fX) = 94,$

$\Sigma(fX^2) = 332,$

$$\sigma^2 = \frac{332}{30} - \left(\frac{94}{30}\right)^2$$

$$= 11{\cdot}0667 - 9{\cdot}8178$$

$$= 1{\cdot}2489.$$

$$\therefore \quad \sigma = \sqrt{1{\cdot}2489} = 1{\cdot}117.$$

3.viii. This method becomes laborious if the figures in the X column are large, since X^2 will be very much larger and the $f(X^2)$ column will require a good deal of calculation and addition. It is usually easiest, especially if N is large and the range of the variable wide, to work from a grouped frequency distribution table. The formulae for the standard deviation in this case are:

$$\sigma_\omega^2 = \frac{\Sigma(fx^2)}{N} - \left\{\frac{\Sigma(fx)}{N}\right\}^2. \tag{7}$$

Using formula (3), this may also be written

$$\sigma_\omega^2 = \frac{\Sigma(fx^2)}{N} - D^2. \tag{7 A}$$

Also

$$\sigma = \sigma_\omega \times \omega. \tag{8}$$

In these formulae σ_ω is the standard deviation in working units. The method of calculation is an extension of that used in Example 3 and the whole process is shown below.

Example 9. Calculate the standard deviation of the data in Table II.

X	f	x	fx	fx^2
190–193	2	−7	−14	98
194–197	4	−6	−24	144
198–201	7	−5	−35	175
202–205	12	−4	−48	192
206–209	19	−3	−57	171
210–213	24	−2	−48	96
214–217	27	−1	−27	27
218–221	35	0	−253	0
222–225	26	1	26	26
226–229	21	2	42	84
230–233	18	3	54	162
234–237	13	4	52	208
238–241	6	5	30	150
242–245	5	6	30	180
246–249	2	7	14	98
250–253	1	8	8	64
	222		256	1875
			−253	
			3	

The fx^2 column is obtained by multiplying together the corresponding entries in the x and the fx columns.

$$\sigma_\omega^2 = \frac{1875}{222} - \left(\frac{3}{222}\right)^2$$

$$= 8{\cdot}4457.$$

$$\therefore \quad \sigma_\omega = \sqrt{8{\cdot}4457} = 2{\cdot}906.$$

$$\therefore \quad \sigma = 2{\cdot}906 \times 4 = 11{\cdot}624.$$

3.ix. The coefficient of variation. If we wish to compare the scatter of different variables about their means, it is useful to be able to express the scatter in some form which is not dependent on the absolute size of the variables. For instance, mice and men may be relatively equally variable in length, but this would not be revealed by stating the standard deviation in inches of a sample of each. Comparison may, however, be usefully made by calculating the *coefficient of variation* in each case. This is readily obtained by expressing the standard deviation as a percentage ratio of the mean, i.e.

$$V = \frac{100\sigma}{m}, \tag{9}$$

where m is the mean of the sample.

The coefficient V is a ratio and is therefore independent of the units in which the mean and standard deviation are measured.

Example 10. Compare the relative variability of the data in Examples 7 and 9.

In Example 7, $\bar{X} = 6,$

$$\sigma = 3{\cdot}16.$$

$$\therefore \quad V = \frac{100 \times 3{\cdot}16}{6} = 52{\cdot}67.$$

In Example 9, $\bar{X} = 219{\cdot}55,$

$$\sigma = 11{\cdot}624.$$

$$\therefore \quad V = \frac{100 \times 11{\cdot}624}{219{\cdot}55} = 5{\cdot}29.$$

Hence the data in Example 7 are about 10 times as scattered relatively to their mean as are those in Example 9.

EXERCISES ON CHAPTER III

6. Find the upper and lower quartiles of the scores in each test from A to G inclusive for the whole 100 subjects and thence derive the semi-inter-quartile range of each test.

7. Calculate the mean deviation of the scores in each of the tests from A to G, using the medians given in the answers to Exercise 4, for

(*a*) subjects 1–50,
(*b*) subjects 51–100.

8. Using formula (5), calculate the standard deviation of the scores in

(*a*) test E, subjects 1–25,
(*b*) test D, subjects 76–100.

9. Calculate the standard deviation of the following test scores, using formula (6); tests *A*, *C*, *D*, *F* and *G*:

(*a*) subjects 1–25,
(*b*) subjects 26–50,
(*c*) subjects 51–75,
(*d*) subjects 76–100.

10. Using the grouped frequency distribution tables constructed in Exercise 3, calculate the standard deviation of the scores of each test from *A* to *G* for the whole 100 subjects. Use formulae (7A) and (8).

From these standard deviations and the means given in the answers to Exercise 3(*a*), calculate the coefficient of variation of each test score. Use formula (9).

Chapter IV

THE NORMAL AND BINOMIAL DISTRIBUTIONS

4.1. Many of the methods of statistics depend on the assumption that the variables under consideration are *normally distributed* and most of the methods cannot strictly be applied unless this assumption is justified. Biological variables are often in point of fact so distributed but it is not safe to assume that any particular distribution is normal without examination.

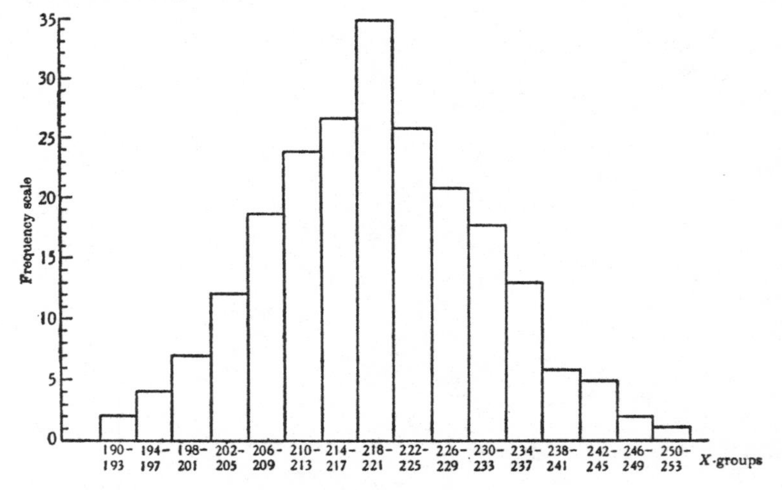

Fig. 1. Histogram of frequency distribution.

The meaning of a normal distribution may be most easily understood by considering certain graphs. Suppose we construct a graph of the data in Table II. Along the base line we first mark off points at equal intervals to represent the X-groups. On the left we make a vertical scale to indicate frequencies. At each point marking an X-group we then draw vertical lines to represent frequencies: the requisite length of these lines is indicated by the frequency of a particular X-group and the frequency scale on the left. Finally we join these vertical lines together by horizontal lines and we have what is called a *histogram* of the frequency distribution of Table II. This is illustrated in Fig. 1.

This figure gives a picture of the distribution of the variable X and it shows that most of the observations are clustered about the middle of the range and that there are relatively few observations at the extremes. The total frequency of the observations is given by the area between the base-line, the top of the histogram and the vertical lines at the boundaries if the interval between the groups is 1 unit. It will be seen that the

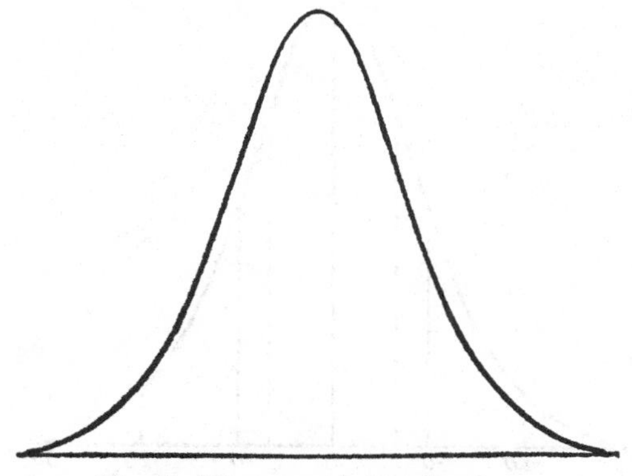

Fig. 2. The normal distribution.

top of the histogram is rather irregular. If, however, a very large number of observations had been made and the points along the base-line had been much more numerous, then the boundary of the histogram would have become a smooth, continuous line, in shape something like the curve shown in Fig. 2.

Fig. 2 shows the shape of a normal distribution, sometimes called a Gaussian distribution. This curve has various important properties, some of which must be mentioned.

4.ii. The normal distribution has an exact mathematical formula. It is a continuous curve and applies to continuous variables, such as height, where the difference between one value of the variable and the next can be indefinitely small.

Mathematically the curve stretches to infinity in both directions, but practically only the portion drawn above is of importance.

The mean of the observations is in the exact centre of the curve and there is the greatest number of observations at this point. Since the curve is symmetrical, the median and the mode coincide with the mean.

The area between the curve and two uprights drawn at any points gives the fraction of the total number of observations between those points. In Fig. 3 uprights have been drawn at points corresponding to 1, 2 and 3 times the standard deviation of the distribution on each side of the mean.

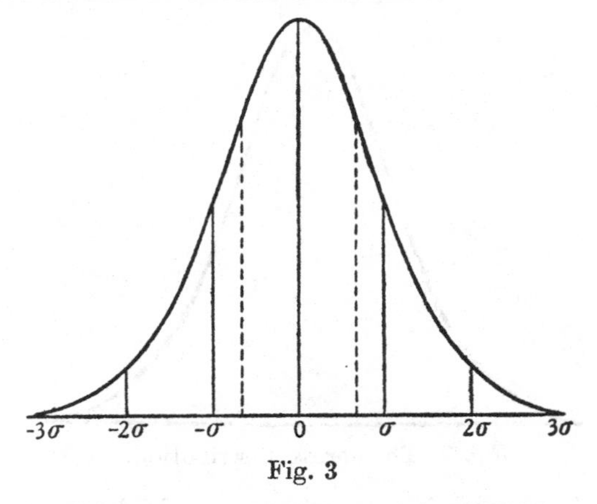

Fig. 3

The area between $-\sigma$ and σ is 68 % of the total area. This means that in a normal distribution 68 % of the observations lie within a distance equal to the standard deviation on each side of the mean. Similarly, from -2σ to 2σ includes 95 %, and from -3σ to 3σ includes 99·7 % of the observations. Hence it is obvious that in a normal distribution practically all the observations lie within a range of 6 times the standard deviation. This provides a rough check on the size of a calculated standard deviation: if the number of observations is large, the standard deviation should be approximately a sixth of the range. (Note then in Example 9 the standard deviation was 11·6 and the range was 63, nearly 6 times as great.)

Exactly half the observations are included in an area bounded by a distance of $0{\cdot}67449\sigma$ on each side of the mean, shown by dotted lines in Fig. 3. This means that the chances are exactly equal that a single observation shall deviate from the mean by an amount greater or less than $0{\cdot}67449\sigma$. Use is made of this fact in calculating *probable errors*, which will be explained later.

Expressing the above measurements in a different way we may say that a value of X whose deviation from the mean, either positively or negatively, is greater than σ will occur roughly 1 in 3 times; a positive or negative deviation greater than 2σ will occur about 1 in 20 times, and greater than 3σ about 1 in 370 times. Tables have been calculated showing the probability of obtaining deviations of any size. Such a table is called a Table of Probability Integral and may be found in *Tables for Statisticians and Biometricians* (Ref. 3).

On such facts is based the conception of statistical *significance*. The term 'significance' is used in statistics to indicate that the odds are heavy against the deviation from its expected value of a particular estimate, difference or coefficient occurring by chance as a result of random sampling. In practice odds of 19 to 1 against an occurrence by chance are usually taken as indicating the significance of that occurrence. This corresponds roughly to the odds of getting a deviation from the mean of a normal distribution greater than twice the standard deviation, either positively or negatively. (Some statisticians prefer heavier odds, such as 99 to 1, as their criterion of significance. This must to some extent depend on the nature of the variables, but in general a probability of 19 to 1 against an occurrence is usually regarded as sufficient for significance.)

Probability, which is symbolised as P, is usually expressed as a decimal fraction, so that odds of 19 to 1 and 99 to 1 are written as $P = 0{\cdot}05$ and $P = 0{\cdot}01$ respectively. Writers sometimes refer to these as the 5 % and 1 % levels of significance.

4.iii. Testing the normality of a distribution. It is unsafe to regard any bell-shaped distribution as being necessarily a normal distribution, and since so much of statistical method depends on normality it is important to be able to test any given distribution for normality. This may be done quite simply by mathematical means, although the process requires a good deal of arithmetic.

Essentially, testing the normality of a distribution depends on the calculation of two constants, β_1 and β_2, which are derived from the first four *moments about the mean* of the distribution, and two further quantities, γ_1 and γ_2, which are related to β_1 and β_2 according to the following equations:

$$\gamma_1 = \pm \sqrt{\beta_1},$$
$$\gamma_2 = \beta_2 - 3.$$

γ_1 is a measure of whether or not the distribution is symmetrical. γ_2 measures departures of a symmetrical nature from normality. The use of these constants will be explained later.

Now the first four moments about the mean of a frequency distribution are denoted by μ_1, μ_2, μ_3 and μ_4, and these are calculated, with certain theoretical corrections, by a method which is simply an extension of that already used in calculating the standard deviation, as in Example 9. The student should now be familiar with the method of calculating $\Sigma(fx)$ and $\Sigma(fx^2)$. Two further columns have to be constructed, the totals of which will yield $\Sigma(fx^3)$ and $\Sigma(fx^4)$. If these four totals are divided by N we get four quantities denoted by ν'_1, ν'_2, ν'_3 and v'_4. The moments about the mean are then obtained from the equations:

$$\mu_1 = 0,$$
$$\mu_2 = \nu'_2 - \nu'^2_1,$$
$$\mu_3 = \nu'_3 - 3\nu'_1\nu'_2 + 2\nu'^3_1,$$
$$\mu_4 = \nu'_4 - 4\nu'_1\nu'_3 + 6\nu'^2_1\nu'_2 - 3\nu'^4_1.$$

When the variate is continuous certain corrections have to be applied for grouping, and the equations then become:

$$\mu_1 = 0,$$
$$\mu_2 = \nu'_2 - \nu'^2_1 - \tfrac{1}{12},$$
$$\mu_3 = \nu'_3 - 3\nu'_1\nu'_2 + 2\nu'^3_1,$$
$$\mu_4 = \nu'_4 - 4\nu'_1\nu'_3 + 6\nu'^2_1\nu'_2 - 3\nu'^4_1 - \tfrac{1}{2}\mu_2 - \tfrac{1}{80}.$$

Having obtained these four moments, β_1 and β_2 are given by the formulae

$$\beta_1 = \frac{\mu_3^2}{\mu_2^3}, \quad \beta_2 = \frac{\mu_4}{\mu_2^2}.$$

From these γ_1 and γ_2 are readily calculated. The *standard errors** of γ_1 and γ_2 are $\sqrt{(6/N)}$ and $\sqrt{(24/N)}$ respectively. This means that if the values of γ_1 and γ_2 are less than *twice* these standard errors, then the distribution is not significantly different from the normal form: if they are greater than twice their standard errors, the distribution is not normal.

This mass of symbolism will probably be alarming to the elementary student, but the actual process of calculation involves arithmetic only and is shown in the following example.

Example 11. Test the normality of the distribution given in the first two columns of the table below:

f	x	fx	fx^2	fx^3	fx^4
1	−5	−5	25	−125	625
2	−4	−8	32	−128	512
5	−3	−15	45	−135	405
10	−2	−20	40	−80	160
20	−1	−20	20	−20	20
50	0	−68	0	−488	0
22	1	22	22	22	22
11	2	22	44	88	176
5	3	15	45	135	405
3	4	12	48	192	768
1	5	5	25	125	625
130		76	346	562	3718
		−68		−488	
		8		74	

$$\nu_1' = 8/130 = 0{\cdot}0615,$$
$$\nu_2' = 346/130 = 2{\cdot}6615,$$
$$\nu_3' = 74/130 = 0{\cdot}5692,$$
$$\nu_4' = 3718/130 = 28{\cdot}6000.$$

* The meaning of the term ‘standard error’ is explained in Chapter v.

Hence

$$\mu_2 = 2{\cdot}6615 - 0{\cdot}0615^2 - 0{\cdot}0833 = 2{\cdot}5744,$$

$$\mu_3 = 0{\cdot}5692 - (3 \times 2{\cdot}6615 \times 0{\cdot}0615) + (2 \times 0{\cdot}0615^3) = 0{\cdot}0787,$$

$$\mu_4 = 28{\cdot}6 - (4 \times 0{\cdot}0615 \times 0{\cdot}5692) + (6 \times 0{\cdot}0615^2 \times 2{\cdot}6615) - (3 \times 0{\cdot}0615^4) - \tfrac{1}{2}(2{\cdot}5744) - 0{\cdot}0125 = 27{\cdot}2207.$$

$$\therefore \quad \beta_1 = \frac{0{\cdot}0787^2}{2{\cdot}5744^3} = 0{\cdot}000363,$$

$$\beta_2 = \frac{27{\cdot}2207}{2{\cdot}5744^2} = 4{\cdot}1072;$$

$$\gamma_1 = 0{\cdot}0191 \quad (\gamma_1 \text{ has the same sign as } \mu_3),$$

$$\text{S.e.} = \sqrt{\frac{6}{130}} = 0{\cdot}215;$$

$$\gamma_2 = 1{\cdot}1072,$$

$$\text{S.e.} = \sqrt{\frac{24}{130}} = 0{\cdot}430.$$

It will be seen that γ_1 is considerably less than twice its standard error, hence the distribution is symmetrical. γ_2, however, is more than twice its standard error, so that the distribution departs from normality in a symmetrical manner. γ_2 is said to measure *kurtosis*. Curves which are flat-topped and short-tailed compared with the normal curve are called *platykurtic*: for these β_2 is less than 3. Curves which are sharply peaked and long-tailed, and for which β_2 is greater than 3, are called *leptokurtic*. In the above example, β_2 is greater than 3, so that the distribution is leptokurtic and not normal. The student should draw a histogram of this distribution and note the peaked shape of it.

4.iv. The above is not the only method of testing the normality of a distribution. An alternative method is to fit a normal curve to a frequency distribution and test the goodness of fit by the χ^2 method. (See Section 9.vii.)

4.v. The binomial distribution and the calculation of probabilities. In some cases the significance of observed results may be tested by use of the binomial distribution. This distribution is well known in algebra and consists of the expansion of the expression $N(p+q)^n$. The full expansion of this general expression, written symbolically, is

$$N(p+q)^n = N\left\{p^n + np^{n-1}q + \frac{n(n-1)}{2!}p^{n-2}q^2 + \frac{n(n-1)(n-2)}{3!}p^{n-3}q^3 + \ldots + q^n\right\}. \quad (10)$$

Usually in probability problems, p is the probability of an event happening, expressed as a fraction, and q the probability of its not happening, so that $p+q=1$. N is the number of *trials* made and n is the number of *events* in each trial. For example, suppose we toss 4 pennies 32 times. Tossing 1 penny is an event; tossing 4 at once is a trial of 4 events and in our experiment we make 32 trials. We wish to know the probabilities of getting 4 heads, 3 heads and 1 tail, etc. Since the chance of getting either a head or a tail in one event is $\frac{1}{2}$, the probabilities of getting the different results will be given by the successive terms in the expansion of $32(\frac{1}{2}+\frac{1}{2})^4$. Substituting these figures in the general expansion (10) we obtain

$$32(\tfrac{1}{2}+\tfrac{1}{2})^4 = 32\left\{(\tfrac{1}{2})^4 + 4(\tfrac{1}{2})^3(\tfrac{1}{2}) + \frac{4.3}{1.2}(\tfrac{1}{2})^2(\tfrac{1}{2})^2 + \frac{4.3.2}{1.2.3}(\tfrac{1}{2})(\tfrac{1}{2})^3 + (\tfrac{1}{2})^4\right\}.$$

Evaluating the successive terms in this we have, calling heads H and tails T,

chances of getting $4H$ and $0T$ are $(\frac{1}{2})^4 \times 32 = 2$

chances of getting $3H$ and $1T$ are $4(\frac{1}{2})^3(\frac{1}{2}) \times 32 = 8$

chances of getting $2H$ and $2T$ are $\frac{4.3}{1.2}(\frac{1}{2})^2(\frac{1}{2})^2 \times 32 = 12$

chances of getting $1H$ and $3T$ are $\frac{4.3.2}{1.2.3}(\frac{1}{2})(\frac{1}{2})^3 \times 32 = 8$

chances of getting $0H$ and $4T$ are $(\frac{1}{2})^4 \times 32 = 2$

Total 32

If we wished to know the odds of getting at least $2H$ showing in each trial, we could find them by summing all the terms involving two or more heads, i.e. the first three terms. Hence out of 32 trials we should expect to see two or more heads in $2+8+12 = 22$ trials. The process is similar in dice-throwing. Suppose, for instance, we throw 3 dice 60 times: how often should we expect to get 3 sixes? The probability p of getting a six in a single event is $\frac{1}{6}$, so that the chances of getting 3 sixes in 60 trials of 3 events is given by the first term in the expansion of $60(\frac{1}{6}+\frac{5}{6})^3$. This term is $60(\frac{1}{6})^3 = \frac{60}{216}$. We should not therefore expect to get 3 sixes in as few as 60 trials, in fact we should need to make at least 216 throws before we could expect this to occur once. This does not mean, of course, that 3 sixes would be thrown at the 216th throw—they might be thrown at the first attempt—but it means that the chances of throwing 3 sixes are 1 in 216, so that in a very large number of trials only 1 in 216 would yield this result.

A more complicated problem would arise if we asked how often we could expect a total score of 15 or more if we threw 3 dice 60 times. First we should need to know how many different scores are possible. The answer to this is obviously 16 since the total score may be any number from 3 to 18 inclusive. However, all these sixteen scores are not equally likely. For instance, a score of 18 can occur in one way only, i.e. 3 sixes, but a score of 17 can occur in three ways, i.e. 6, 6, 5 or 6, 5, 6 or 5, 6, 6. A score of 16 may occur in 6 ways, viz. 6, 6, 4 : 6, 4, 6 : 4, 6, 6 : 6, 5, 5 : 5, 6, 5 or 5, 5, 6. Working out all the possibilities in this way we arrive at the following table:

Score	No. of possible ways	Score	No. of possible ways
18	1	10	27
17	3	9	25
16	6	8	21
15	10	7	15
14	15	6	10
13	21	5	6
12	25	4	3
11	27	3	1
		Total	216

By addition we see that there are 216 ways of getting different scores. As a check on this total, note that there are 6 ways in which the first die may fall, and for each of these there are 6 ways in which the second die may fall, and for each combination of the first and second there are 6 ways in which the third may fall. The total number of possible combinations of the three, therefore, is $6 \times 6 \times 6 = 216$. Of these 216 possible score combinations, $1+3+6+10 = 20$ will give a score of 15 or more, hence the number of times we should expect a score of 15 or more in 60 throws of 3 dice is

$$60 \times \frac{20}{216} = 5{\cdot}5.$$

Although this illustration does not directly make use of the binomial expansion, it gives an example of the calculation of probabilities—a process which is not always obvious.

As a further example, using the binomial expansion, take the case of a football coupon where the results of 14 matches have to be forecast. The result may be a win, a draw or a loss in any one match. Assuming that the chances of a win, a draw or a loss are equal (this is probably not true in practice but it is assumed here for the sake of illustration), the probability of getting any single forecast correct by chance is 1 in 3. What is the probability of getting all 14 forecasts correct and all 14 wrong? The answer to this is given by the first and last terms of the expansion of the binomial $(\frac{1}{3}+\frac{2}{3})^{14}$. The first term is $(\frac{1}{3})^{14} = 1/4{,}782{,}969$. The last term is

$$(\tfrac{2}{3})^{14} = 16{,}384/4{,}782{,}969 = 1/342{\cdot}55.$$

Hence to make sure of one correct forecast in all 14 cases one would have to fill in over $4\frac{3}{4}$ million coupons, each with a different forecast, whereas on the average 1 form in every 343 would be completely wrong.

4.vi. Now the mean of a binomial distribution $(p+q)^n$ is np and the standard deviation is $\sqrt{(npq)}$. Suppose we toss a penny 100 times and count a head as a success. If the penny is unbiased, the mean number of successes would be $100 \times \frac{1}{2} = 50$. If, however, in a particular trial we obtained 62 heads, could we regard the penny as being biased? The standard deviation is $\sqrt{(100 \times \frac{1}{2} \times \frac{1}{2})} = 5$. The observed value of 62 successes differs from the expected mean by 12, and 12 is $2{\cdot}4$ times the standard deviation. Reference to the table of the probability integral shows that the probability of getting a deviation from the mean of $2{\cdot}4\sigma$ or greater is just less than 0·02, so that we should strongly suspect the penny of being biased. Note that in this example we should have obtained the same result if we had had 38 observed successes, since the probability of a deviation of $2{\cdot}4\sigma$ or more on *either* side of the mean is just less than 0·02. If, however, the penny were tossed in an electrical field which we had reason to suspect would cause more heads to appear, we should have to ask what is the probability of obtaining the same deviation from the mean in a *positive* direction only. This would be *half* that given in the table, i.e. with 62 heads it would be just less than 0·01 (0·0082 to be exact). Hence we should conclude that the electrical field was definitely having the suspected effect.

The question of when to use P and when to use $\frac{1}{2}P$ for testing significance is apt to be a difficult one. The answer depends entirely upon the hypothesis we are testing. For example, a test is given to 100 subjects twice and it is found that 59 subjects get a better score the second time and 41 get a worse. With these data we may test two hypotheses.

(*a*) Assuming that it is equally likely for a subject to get a better or a worse score the second time, the chance of a subject getting a better score is $\frac{1}{2}$. We may then test the hypothesis that the observed distribution of better and worse scores does not differ significantly from chance. As in the previous example, we should expect 50 subjects to improve by chance with a standard deviation of 5. The observed deviation from the mean is 9 which is 1·8 times σ. From the tables it may be seen that for this deviation $P = 0{\cdot}071$. Hence we conclude that the observed distribution does not differ significantly from chance.

(*b*) We may, however, decide on psychological grounds that doing a test twice is likely to have a disturbing influence on the scores, and we proceed to test the hypothesis that twice-testing will cause an improvement in scores. Here we are dealing with deviations from the mean in a *positive* direction only, so that the probability is $\frac{1}{2}P = 0{\cdot}035$. From this we should conclude that scoring on the test has improved significantly on a second testing. Note that this shows only that a greater number of individuals than would be expected by chance improve on a second testing. The *amount* of improvement, as measured by the number of marks scored, might still not be significant. To decide this further point it would be necessary to average the 100 differences between the first and second scores and apply the t test to see if this mean differed significantly from zero.

As a note of warning, it must be emphasised that the grounds for using $\frac{1}{2}P$ as a measure of significance must be exceedingly firm and justified *before* making the actual test. If there is the slightest doubt, it is far safer in the interests of truth to use P as a criterion. This is a matter of statistical ethics. The honest inquirer makes up his mind before analysing his data what can reasonably be expected from them, and for conviction about the truth of his statistical results he must feel that they are not based on any specious argument, however plausible.

EXERCISES ON CHAPTER IV

11. Using the method of Section 4.iii (Example 11), test the normality of the distribution of the scores in test A for the whole 100 subjects. Make use of the grouped frequency distribution table constructed in Exercise 3.

12. Repeat Exercise 11 for the whole 100 scores in test C.

13. Five pennies are tossed 320 times. How often would you expect at least 4 heads?

14. Sixty-four people are asked a problem question, the answer to which can be only 'Yes' or 'No'; 38 people answer 'Yes', which is correct, and 26 answer 'No'. What are the chances of obtaining this result if all the answers were guessing?

Chapter V

SIGNIFICANCE OF MEAN AND DIFFERENCE BETWEEN MEANS

5.i. The observations recorded in a single biological experiment are but one sample drawn from the whole population of possible samples. If a second experiment is made it is unlikely that the mean of the observations in this case will be identical with that of the first experiment. In short, it will be found that a large number of experiments will yield many different values of the mean, each one departing more or less from the true mean of the whole population.

If the standard deviation of the whole population is σ_p and we take a large number of random samples of n observations, then the means of the sample will be distributed with a standard deviation $\sigma_p/\sqrt{n}$. If the population is normally distributed, the means also will be normally distributed. Even if the distribution of the population is not normal, the distribution of the means of samples still tends to be normal if the size of the samples is sufficiently large, but in the case of small samples the distribution of the means is not normal.

Usually we do not know the standard deviation of the whole population but have to take the standard deviation of an observed sample as an estimate of it. In this case we estimate the standard deviation of the sampling distribution from the number and standard deviation of a single sample. This estimated value is called the *standard error of the mean*, i.e.

$$\text{S.e. of mean} = \frac{\sigma}{\sqrt{N}}, \tag{11}$$

where σ is the standard deviation of the sample and N the number of observations in it.

(There was formerly a practice, which has little to recommend it, of calculating the *probable error* of the mean. This is given by the formula

$$\text{P.e. of mean} = \frac{0{\cdot}67449\sigma}{\sqrt{N}}. \tag{11 A}$$

It may be noted that three times the probable error is roughly equal to twice the standard error.)

5.ii. Significance of a single mean. If we have only a single sample to give estimates of $\bar{X}$ and σ, then the distribution of $\bar{X}/\sigma$ will not be normal. However, the correct distribution has been calculated and tables have been made enabling us to make use of the data from a single sample; an extract from these tables will be given later.

When calculating the standard deviation for the purpose of examining the significance of the mean, $S(X-\bar{X})^2$ should be divided by $(N-1)$ instead of by N. The reason for this is that we are making an *estimate* of the standard deviation in the whole population and this is best obtained by dividing the sum of the squared deviations from the mean by the number of *degrees of freedom* available. This number is one less than the number in the sample.

TABLE III. *Values of t corresponding to a probability* $P = 0{\cdot}05$

n	t	n	t	n	t
1	12·706	11	2·201	21	2·080
2	4·303	12	2·179	22	2·074
3	3·182	13	2·160	23	2·069
4	2·776	14	2·145	24	2·064
5	2·571	15	2·131	25	2·060
6	2·447	16	2·120	26	2·056
7	2·365	17	2·110	27	2·052
8	2·306	18	2·101	28	2·048
9	2·262	19	2·093	29	2·045
10	2·228	20	2·086	30	2·042

For $n = \infty$, $t = 1{\cdot}96$.

The above table is an extract from a full table given by R. A. Fisher (Ref. 2), Table IV; or in the Fisher and Yates Tables (Ref. 5), Table III.

Having obtained the mean and standard deviation we need to calculate a statistic known as t; this is essentially the ratio of the mean to its standard error. (t may also be the ratio of a difference between means to its standard error: see below, Section 5.iii (*b*).)

For a single mean

$$t = \bar{X} \div \frac{\sigma}{\sqrt{N}} = \frac{\bar{X}\sqrt{N}}{\sigma}. \tag{12}$$

In Table III are given values of t corresponding to different values of n, the number of degrees of freedom, i.e. $n = N-1$ in this case. The odds against values of t as big as or bigger than these occurring by chance are 19 : 1, i.e. the *probability*, usually denoted by P, of their occurring by chance is 0·05. If the calculated value of t is greater than that given in the table for the appropriate value of n, then the mean is significantly different from zero.

Example 12. Ten schoolchildren were given an arithmetic test. They were then given a month's further tuition and a second test of equal difficulty was held at the end of it. Their marks in these two tests are given below.

Scholar	Test 1	Test 2
1	20	22
2	18	19
3	19	17
4	22	18
5	17	21
6	20	23
7	19	19
8	16	20
9	21	22
10	19	20

Do these marks give evidence that the scholars had benefited by the extra tuition?

This problem resolves itself into the question, Is the mean of the differences between successive marks significantly different from zero?

We need first to construct a third column giving the values of (Test 2 – Test 1); this will be our X column. Add this column to get $S(X)$ and obtain $\bar{X}$ by dividing $S(X)$ by N (formula (1)). Make a fourth column giving $(X-\bar{X})$ and a fifth giving $(X-\bar{X})^2$. Add this fifth column to obtain $S(X-\bar{X})^2$. This gives us all the necessary data for calculating t. The last three columns and the remainder of the working are shown below.

Test 2 – Test 1

X	$X-\bar{X}$	$(X-\bar{X})^2$
2	1	1
1	0	0
−2	−3	9
−4	−5	25
4	3	9
3	2	4
0	−1	1
4	3	9
1	0	0
1	0	0
10		58

$$S(X) = 10,$$
$$\bar{X} = 1,$$
$$\frac{S(X-\bar{X})^2}{N-1} = \frac{58}{9}.$$
$$\therefore \quad \sigma = \sqrt{\frac{58}{9}}.$$

Hence, by formula (12),

$$t = \frac{1 \times \sqrt{10}}{\sqrt{\frac{58}{9}}} = \sqrt{\frac{90}{58}}$$
$$= 1{\cdot}25.$$

Reference to Table III shows that for $n = 9$, $t = 2{\cdot}262$. Our calculated value is less than this, hence the mean of X is not significantly different from zero, and the marks are insufficient to prove the benefit of the extra tuition.

The above method of calculation is not convenient if the mean is not a whole number. We may, therefore, adopt an alternative method of calculation, making use of the identity

$$S(X-\bar{X})^2 = S(X^2) - \frac{1}{N}[S(X)]^2.$$

In the above example all that is needed is to construct and add a column giving the squares of all the observations, i.e. a column of X^2 yielding a total of $S(X^2)$. In Example 12,

$$S(X^2) = 4+1+4+16+16+9+0+16+1+1 = 68.$$

Hence
$$S(X-\bar{X})^2 = 68 - \tfrac{1}{10}(10)^2$$
$$= 68 - 10$$
$$= 58.$$

From this point the calculation of t is identical with that given in Example 12.

5.iii. Significance of the difference between means. An important and often occurring problem is to determine whether there is a real difference between two observational means or not. In statistical language this problem may be expressed in the words, Is the difference between the means such that they might have been drawn from the same population by random sampling or are they drawn from two different populations? There are two methods of dealing with this question appropriate to the cases in which the samples are large or in which they are small.

(*a*) *Large samples.* If the numbers of observations in the two samples are large, say, at least 50, the question may be settled by calculating the *standard error of the difference* between the means. If the means are $\bar{X}_1$ and $\bar{X}_2$, their standard deviations σ_1 and σ_2 and the numbers in the samples N_1 and N_2 respectively, then the standard error of the difference between the means is given by the formula

$$\text{S.e. of difference} = \sqrt{\left(\frac{\sigma_1^2}{N_1} + \frac{\sigma_2^2}{N_2}\right)}. \qquad (13)$$

This formula applies only if the two variables are independent or uncorrelated (see Chapter VI).

If the two variables are correlated

$$(\text{S.e. of difference})^2 = \frac{\sigma_1^2}{N_1} + \frac{\sigma_2^2}{N_2} - 2r\frac{\sigma_1\sigma_2}{\sqrt{(N_1N_2)}},$$

where r is the coefficient of correlation.

As before, the *probable error* of the difference would be

$$\text{P.e. of difference} = 0{\cdot}67449\sqrt{\left(\frac{\sigma_1^2}{N_1} + \frac{\sigma_2^2}{N_2}\right)}. \qquad (13\,\text{A})$$

If the difference between the two means is greater than twice its standard error (or three times its probable error), then the means are significantly different, i.e. it is unlikely that they would be drawn from the same population by random sampling, the odds against being at least 19 to 1.

Example 13. A group of boys and a group of girls were given an intelligence test. The mean scores, standard deviations and numbers in the groups were as follows:

	Boys	Girls
Mean	124	121
σ	12	10
N	72	50

Was the mean test score of the boys significantly greater than that of the girls?

$$\text{Difference between the means} = 124 - 121 = 3,$$

$$\text{S.e. of difference} = \sqrt{\left(\frac{144}{72} + \frac{100}{50}\right)} = 2.$$

Hence
$$\frac{\text{Difference}}{\text{S.e. of difference}} = \frac{3}{2} = 1\cdot5.$$

In this experiment, therefore, the mean intelligence test score of the boys was not significantly greater than that of the girls.

(*b*) *Small samples.* When we wish to compare the means of small samples of less than 50 observations, the use of formula (13) is no longer a sufficiently strict test and we have to apply the t test of significance. The test is essentially similar to that in the previous section but corrections have to be made to allow for sampling errors, which are more important in small samples.

Suppose we have N_1 readings of a variable x_1 and N_2 readings of a variable x_2 and we wish to see whether or not the means $\bar{x}_1$ and $\bar{x}_2$ differ significantly from one another (or, in other words, we wish to see what is the probability that the two samples could be drawn from the same population). To apply the t test in this case we need to know six quantities, N_1, N_2, $S(x_1)$, $S(x_2)$, $S(x_1^2)$ and $S(x_2^2)$.

As usual,
$$\bar{x}_1 = \frac{S(x_1)}{N_1} \quad \text{and} \quad \bar{x}_2 = \frac{S(x_2)}{N_2}.$$

The variance of the combined observations, which we shall denote by σ_d^2, is

$$\sigma_d^2 = \frac{S(x_1 - \bar{x}_1)^2 + S(x_2 - \bar{x}_2)^2}{N_1 + N_2 - 2},$$

and the standard error of the difference is

$$\text{S.e. of difference} = \sigma_d \sqrt{\left(\frac{1}{N_1} + \frac{1}{N_2}\right)}.$$

(The reader should compare this expression with formula (13) in the case where $\sigma_1 = \sigma_2$.)

In this case t is the ratio of the difference between the means to the standard error of the difference, i.e.

$$t = \frac{|\bar{x}_1 - \bar{x}_2|}{\sigma_d \sqrt{\left(\frac{1}{N_1} + \frac{1}{N_2}\right)}}. \tag{14}$$

We must now express σ_d in terms of the six quantities with which we started. We have

$$\sigma_d^2 = \frac{S(x_1^2) - \frac{[S(x_1)]^2}{N_1} + S(x_2^2) - \frac{[S(x_2)]^2}{N_2}}{N_1 + N_2 - 2}.$$

Written in full, therefore,

S.e. of difference

$$= \sqrt{\left\{\left(\frac{1}{N_1} + \frac{1}{N_2}\right) \frac{1}{N_1 + N_2 - 2}\right\}}$$
$$\times \sqrt{\left\{S(x_1^2) - \frac{[S(x_1)]^2}{N_1} + S(x_2^2) - \frac{[S(x_2)]^2}{N_2}\right\}}.$$

The part dealing with N_1 and N_2 under the first square-root sign reduces to

$$\sqrt{\left(\frac{N_1 + N_2}{N_1 . N_2 . (N_1 + N_2 - 2)}\right)},$$

for which we may write $\sqrt{N'}$. For the convenience of students, values of $\sqrt{(1/N')}$ for N_1 and N_2 between 10 and 50 have been tabulated in Appendix A.

Using therefore the six quantities with which we started, we may therefore express t in the following form:

$$t = \frac{|\bar{x}_1 - \bar{x}_2| \sqrt{\frac{1}{N'}}}{\sqrt{\left\{S(x_1^2) - \frac{[S(x_1)]^2}{N_1} + S(x_2^2) - \frac{[S(x_2)]^2}{N_2}\right\}}}. \tag{14 A}$$

The use of this formula is shown in the example below.

In this case, for using Table III to find the probability of obtaining observed values of t, the value of n is

$$n = N_1 + N_2 - 2.$$

Note. The t test is applicable only when the variates are normally distributed and are not correlated.

Example 14. The span of apprehension of two small groups of children, one from the lowest class in a school and the other from the top class, was tested by seeing how many digits they could repeat backwards from memory after hearing them once repeated forwards. The numbers of digits correctly repeated in the two cases were as follows:

Group A	3	5	6	4	3	3	4	
Group B	5	8	9	6	12	9	7	6

Is there any real difference in the span of apprehension of the two groups?

We have the following data:

For group A, $N_1 = 7$, $S(X_1) = 28$, $S(X_1^2) = 120$, $\bar{X}_1 = 4{\cdot}00$.

For group B, $N_2 = 8$, $S(X_2) = 62$, $S(X_2^2) = 516$, $\bar{X}_2 = 7{\cdot}75$.

Hence $$\bar{X}_2 - \bar{X}_1 = 7{\cdot}75 - 4{\cdot}00 = 3{\cdot}75.$$

Now $$\sqrt{\frac{1}{N'}} = \sqrt{\frac{7 \times 8 \times 13}{15}} = 6{\cdot}97.$$

Also $$S(X_1^2) - \frac{[S(X_1)]^2}{N_1} + S(X_2^2) - \frac{[S(X_2)]^2}{N_2}$$
$$= 120 - \frac{784}{7} + 516 - \frac{3844}{8}$$
$$= 120 - 112 + 516 - 480{\cdot}5$$
$$= 43{\cdot}5.$$

Hence, by formula (14 A),

$$t = \frac{3{\cdot}75 \times 6{\cdot}97}{\sqrt{43{\cdot}5}} = \frac{26{\cdot}1375}{6{\cdot}595}$$
$$= 3{\cdot}96.$$

For using Table III in this case, the value of n for entry is $7+8-2 = 13$. The value of t in Table III corresponding to $n = 13$ is 2·160. Our calculated value of t is considerably greater than this, hence the difference between the means is significant and we conclude that the span of apprehension of group B was significantly greater than that of group A.

If the actual observations in the two groups for comparison are large, the arithmetic in calculating the standard error of the difference between the means may be fairly arduous. In such cases the work may be lightened by working from arbitrary origins, provided the range of the readings is not too great. To do this, we choose a convenient arbitrary origin, or starting point, near the middle of the range of observations. Calling this origin A, we construct a column headed $(X-A)$ by subtracting A from each reading in turn. Care must be taken of the signs since some of the entries will be negative. Summing this column gives us $S(X-A)$. Each entry in this column is then squared and the resulting $(X-A)^2$ column summed to obtain $S(X-A)^2$.

We then make use of the following identities:

$$S(X) = S(X-A)+NA,$$

$$S(X^2) = S(X-A)^2-NA^2+2A\,.\,S(X),$$

$$S(X-\bar{X})^2 = S(X-A)^2-\frac{1}{N}[S(X-A)]^2.$$

The method of extending this process to two groups is illustrated in the following example.

Example 15. Is the difference between the mean reaction times of the following two groups significant?

Group A	98	97	104	106	100	111	99	99	101	102	
Group B	100	94	93	99	101	87	86	91	85	86	89

The range of observations in group A is from 97 to 111, so we may choose 105 as a convenient origin, i.e. $A_1 = 105$. Call the readings in group A, X_1. We then construct a column headed X_1-A_1 and sum it (see table on p. 45). Each reading in this column is then squared, giving a column headed $(X_1-A_1)^2$ which we also sum. The process is repeated with group B taking $A_2 = 90$.

Substituting these totals in the identities given in the table we have for group A,

$$\begin{aligned} S(X_1) &= S(X_1-A_1)+N_1A_1 \\ &= -33+(10\times 105) \\ &= 1017, \end{aligned}$$

whence $\bar{X}_1 = 1017/10 = 101{\cdot}7,$

$$S(X_1-\bar{X}_1)^2 = S(X_1-A_1)^2 - \frac{1}{N}[S(X_1-A_1)]^2$$
$$= 273 - 33^2/10$$
$$= 273 - 108{\cdot}9 = 164{\cdot}1.$$

Similarly for group B,

$$\bar{X}_2 = 91{\cdot}9091,$$
$$S(X_2-\bar{X}_2)^2 = 354{\cdot}9091.$$

X_1-A_1	$(X_1-A_1)^2$	X_2-A_2	$(X_2-A_2)^2$
−7	49	10	100
−8	64	4	16
−1	1	3	9
1	1	9	81
−5	25	11	121
6	36	−3	9
−6	36	−4	16
−6	36	1	1
−4	16	−5	25
−3	9	−4	16
		−1	1
−33	273	21	395

From Appendix A, for $N_1 = 10$ and $N_2 = 11$, we find that $\sqrt{\frac{1}{N'}} = 9{\cdot}98$. Substituting in formula (14 A) we obtain

$$t = \frac{(101{\cdot}7 - 91{\cdot}9091)\,9{\cdot}98}{\sqrt{(164{\cdot}1 + 354{\cdot}9091)}} = \frac{97{\cdot}7132}{22{\cdot}78} = 4{\cdot}3.$$

From Table III with $n = N_1 + N_2 - 2 = 19$, we find a critical value for t of 2·093. Our calculated value is much larger than this, hence the difference between the means of the two groups may be taken as significant.

5.iv. Variance differences: the variance ratio. A point that is frequently lost sight of is that when two means are judged to be significantly different by the t test, this result may be due not to the fact that the two groups of observations are drawn from populations with different means, but to the fact that the variances of the two groups are significantly different. This may, and should, be examined by calculating the *variance ratio* of the two groups and testing its significance by reference to Fisher and Yates's Table V. The variance ratio (V.R.) is simply obtained by working out the variance of each group and dividing the larger by the smaller. The number of degrees of freedom available for this is $N-1$ for each group, but note that in referring to Fisher and Yates's table, n_1 is the number of degrees of freedom in the *larger* variance.

Applying this to the Example 15 of the previous section, we have for the variance of group A, 164·1/9 = 18·23 and for the variance of group B, 354·9091/10 = 35·49.

Hence the V.R. = 35·49/18·23 = 1·95.

Referring to Fisher and Yates's table, we find that for $n_1 = 10$ (since the variance of group B is the larger) and $n_2 = 9$, the critical value of the V.R. is about 3·1 (its exact value could be found by interpolation). The calculated V.R. is much smaller than this, so we conclude that the two variances are not significantly different.

5.v. Significance of difference between proportions. It frequently happens that no exact measurement of a quality is possible but the presence or absence of the quality may be observed. For example, we may note whether or not a person has blue eyes without attempting to measure intensity of blueness. In the same way we may note what proportion of a group of objects have a certain quality and we may wish on occasion to examine whether the proportion of one group possessing a quality is really different from the proportion of another group possessing that same quality. As in the case of the difference between means, the problem may be expressed in the form: What are the odds against obtaining by chance a difference of proportion as big as the one observed in a homogeneous population? Now in the case of investigating the significance of the difference between means, the probability of obtaining differences of various sizes depends on the distribution of the variables, assumed in this instance to be a normal distribution. In the case of proportions the probabilities depend upon the binomial distribution (see Chapter IV).

The parameters of this type of distribution are different from those of a normal distribution but the method of examining the significance of differences of proportions is similar.

Let N be the number of individuals in a group and p be the proportion of them possessing some particular quality. Usually p is expressed as a decimal fraction, and we call the proportion *not* possessing the quality q, so then $q = 1-p$. The standard error of a single proportion is $\sqrt{\frac{pq}{N}}$. If we have two groups of N_1 and N_2 individuals respectively and if the proportions of them possessing the same quality are p_1 and p_2, we need an expression giving us the standard error of the difference between p_1 and p_2.

First we have to estimate the proportion of individuals in the combined groups who possess the particular quality. If p is this proportion, then $p = \frac{p_1 N_1 + p_2 N_2}{N_1 + N_2}$. As before, $q = 1-p$. We have, then

Standard error of difference between proportions

$$= \sqrt{\left\{pq\left(\frac{1}{N_1}+\frac{1}{N_2}\right)\right\}}. \quad (15)$$

The differences between proportions drawn from a homogeneous population are not distributed in the same way as the differences between means drawn from a normal population. It is therefore safer to require an observed difference between proportions to be three times its standard error before assuming significance.

Note. In examining the significance of the difference between proportions, the actually observed proportions must be used. If the proportions are expressed as percentages, the groups concerned are assumed to be of equal size, and this will lead to error except when the groups are actually equal.

Example 16. In a group of 50 boys, 24 are over 4 ft. in height, whilst 18 out of a group of 60 girls are over 4 ft. Can the difference between the proportions of taller children in the two groups be regarded as real?

We have

$$N_1 = 50, \quad p_1 = 24/50 = 0{\cdot}48,$$
$$N_2 = 60, \quad p_2 = 18/60 = 0{\cdot}30.$$

The combined proportion,

$$p = \frac{(0{\cdot}48 \times 50) + (0{\cdot}30 \times 60)}{50 + 60} = \frac{24 + 18}{110} = 0{\cdot}38,$$
$$q = 1 - 0{\cdot}38 = 0{\cdot}62.$$

Hence S.E. of difference of proportions

$$= \sqrt{\left\{pq\left(\frac{1}{N_1} + \frac{1}{N_2}\right)\right\}}$$
$$= \sqrt{\left\{0{\cdot}38 \times 0{\cdot}62\left(\frac{1}{50} + \frac{1}{60}\right)\right\}}$$
$$= \sqrt{(0{\cdot}2356 \times 0{\cdot}03667)}$$
$$= \sqrt{0{\cdot}008640}$$
$$= 0{\cdot}093.$$

The difference

$$p_1 - p_2 = 0{\cdot}48 - 0{\cdot}30 = 0{\cdot}18.$$

The difference is not quite twice its standard error and we therefore conclude that the difference between the proportions of taller children in the two groups cannot be regarded as a real one.

EXERCISES ON CHAPTER V

15. Subtract the score in test D from the score in test C for each of the first 25 subjects. Calculate the mean difference between these scores and determine whether this mean is significantly different from zero or not. Use formula (12).

16. Using formula (13) and the standard deviations given in the answers to Exercise 10, calculate the standard error of the difference between the mean scores of the whole 100 subjects in the following tests:

(*a*) A and G, (*b*) C and D, (*c*) C and F,
(*d*) D and F, (*e*) D and G, (*f*) F and G.

Thence determine which pairs of means are significantly different. Use the means given in the answer to Exercise 3 (*a*).

17. Using the method of Section 5.iii (*b*), formula (14A), examine the significance of the difference between the following pairs of means:

(*a*) Test *A*: mean of subjects 1–25 and 26–50.
(*b*) Test *C*: mean of subjects 1–25 and 26–50.
(*c*) Test *D*: mean of subjects 1–25 and 26–50.
(*d*) Test *G*: mean of subjects 1–25 and 26–50.
(*e*) Mean of subjects 1–25 in test *C* and mean of subjects 1–25 in test *G*.
(*f*) Mean of subjects 51–75 in test *C* and mean of subjects 51–75 in test *G*.
(*g*) Mean of subjects 51–75 in test *A* and mean of subjects 51–75 in test *D*.
(*h*) Mean of subjects 76–100 in test *F* and mean of subjects 76–100 in test *G*.

18. Calculate the variance ratio between the two groups in Example 14. Does this invalidate the conclusion that the means of the two groups are significantly different?

19. In a school of 320 boys in a certain town, 45 % were absent through influenza in one month. In the same month in another town, 39 boys from a school of 150 were absent for the same reason. From these samples, would you regard the prevalence of influenza as being equal in the two towns?

Chapter VI

CORRELATION

6.i. It frequently happens in experimental work that we wish to know the association between two variables, that is, to know to what extent one variable is related to the other. There are various methods of measuring such association, dependent on the nature of the variables, their types of distribution, etc. When the two variables are *numerical* and *normally distributed*, the association, or *correlation*, between them may best be measured by a method known as the *product-moment* method. This is by far the most useful and theoretically satisfactory method of measuring correlation and much advanced statistical work is based upon product-moment correlation. It will therefore be considered first.

6.ii. In describing methods of correlation the two variables will be called X and Y. These variables will have means $\bar{X}$ and $\bar{Y}$ and standard deviations σ_x and σ_y. Since the various values of the variables will always be considered in pairs, an X with a Y, there will be the same number of X's and Y's in any particular case, i.e. N. In terms of these statistics a quantity known as the *coefficient of correlation* may be calculated: this coefficient is denoted by r. If there is complete positive correlation between X and Y, r has the value 1; if there is complete negative correlation it has the value -1, and incomplete correlation gives decimal values for r between 1 and -1. If there is no relation at all between the variables, r is 0.

The meaning of the above statements may be illustrated in this way. Suppose the heights and weights of N people were measured: these would be our X and Y and there would be one value of X and one value of Y relating to each person. Now suppose the tallest person was also the heaviest, the second tallest the second heaviest and so on, until we reached the shortest person who was also the lightest: in this case there would be complete positive association between X and Y and r would be 1 (provided the relation between X and Y was exactly linear and could be expressed by the equation $Y = a + bX$). If, on the other hand, the group of persons was so peculiar that the tallest person was also the lightest, the second tallest the second lightest and so on to the shortest, who would be the heaviest in this case, then there would be complete negative correlation between height and weight, and r would be -1 (with the same proviso as before). Complete correlation is very rare. Usually there is a general but not complete agreement between two variables, so that r is fractional.

6.iii. The general formula for r is quite simple. If $(X - \bar{X})$ and $(Y - \bar{Y})$ are the deviations of corresponding values of X and Y from their means (i.e. the pair of values corresponding to one person), then these two deviations may be multiplied together to give the product $(X - \bar{X})(Y - \bar{Y})$. If we add together all such products for the N persons, we obtain $S(X - \bar{X})(Y - \bar{Y})$. The coefficient of correlation is then given by the formula

$$r = \frac{S(X - \bar{X})(Y - \bar{Y})}{N\sigma_x\sigma_y}. \tag{16}$$

The application of the name 'product-moment' to this method may now be appreciated. The average deviation of any value of X from the mean may be described as the first moment about the mean, and similarly for the Y's. The mean product of such deviations is similarly called a 'product-moment'. The expression $S(X - \bar{X})(Y - \bar{Y})/N$ is called the *co-variance*.

Crude test scores may be transformed into *standard scores* by expressing them as deviations from their mean in terms of their standard deviation. Thus a crude score of X yields a standard score of $\frac{X - \bar{X}}{\sigma_x}$ and a crude score of Y yields a standard score of $\frac{Y - \bar{Y}}{\sigma_y}$. Call these X_s and Y_s respectively. Then $r = \frac{S(X_s Y_s)}{N}$.

Formula (16) is the simple theoretical form. In practice considerations of ease in calculation necessitate modifications of this and we shall now consider some of them.

6.iv. Product-moment correlation when N is small. If we have a small number of X's and Y's, say less than 80, the calculation of r would be performed as follows. Write down the values of X and Y in two parallel columns in their pairs, i.e. so that the pair of readings in each horizontal row belongs to the same person. Next calculate two more columns, headed X^2 and Y^2, by squaring the terms in the first two columns. The totals of these four columns will give us $S(X)$, $S(Y)$, $S(X^2)$ and $S(Y^2)$, and these data will enable us to calculate the means and standard deviations of X and Y by formulae (1) and (5B).

Now instead of subtracting the mean $\bar{X}$ from each value of X and $\bar{Y}$ from each value of Y and multiplying the answers together, we shall get the total of products, or product-sum, by a method similar to that used in Section 3.vii for calculating the standard deviation. We shall multiply together the actual values of X and Y as they stand in columns 1 and 2, sum the products, and then subtract the product of $\bar{X}$ and $\bar{Y}$ at the end, after having divided the product-sum by N. The validity of this may readily be seen by the following simple algebraical proof:

$$(X-\bar{X})(Y-\bar{Y}) = XY - \bar{X}Y - \bar{Y}X + \bar{X}\bar{Y}.$$

$$\therefore \quad S(X-\bar{X})(Y-\bar{Y}) = S(XY) - \bar{X}S(Y) - \bar{Y}S(X) + N\bar{X}\bar{Y}.$$

$$\therefore \quad \frac{S(X-\bar{X})(Y-\bar{Y})}{N} = \frac{S(XY)}{N} - \bar{X}\frac{S(Y)}{N} - \bar{Y}\frac{S(X)}{N} + \bar{X}\bar{Y}$$

$$= \frac{S(XY)}{N} - \bar{X}\bar{Y} - \bar{Y}\bar{X} + \bar{X}\bar{Y}$$

$$= \frac{S(XY)}{N} - \bar{X}\bar{Y}.$$

Accordingly we form a fifth column, headed XY, by multiplying together corresponding X's and Y's in the first two columns. The total of this column is $S(XY)$. We may then obtain r from the formula

$$r = \frac{\frac{S(XY)}{N} - \overline{X}\overline{Y}}{\sigma_x \sigma_y}. \quad (16\,\text{A})$$

Example 17. Twenty pupils are given small tests in arithmetic and Latin, and the marks gained in each test, from a maximum of 10, are shown below.

Pupil	*A*	*B*	*C*	*D*	*E*	*F*	*G*	*H*	*I*	*J*
Arith.	3	9	7	8	4	1	6	9	7	8
Latin	1	8	4	10	6	5	5	3	8	7

Pupil	*K*	*L*	*M*	*N*	*O*	*P*	*Q*	*R*	*S*	*T*
Arith.	5	4	6	5	2	6	5	4	6	2
Latin	2	6	5	9	4	5	7	1	3	5

Calculate the coefficient of correlation between the two sets of marks.

We construct the five columns, as described above, and obtain $S(X)$, $S(Y)$, $S(X^2)$, $S(Y^2)$ and $S(XY)$. The actual arithmetic is shown.

X	Y	X^2	Y^2	XY
3	1	9	1	3
9	8	81	64	72
7	4	49	16	28
8	10	64	100	80
4	6	16	36	24
1	5	1	25	5
6	5	36	25	30
9	3	81	9	27
7	8	49	64	56
8	7	64	49	56
5	2	25	4	10
4	6	16	36	24
6	5	36	25	30
5	9	25	81	45
2	4	4	16	8
6	5	36	25	30
5	7	25	49	35
4	1	16	1	4
6	3	36	9	18
2	5	4	25	10
107	104	673	660	595

$$\overline{X} = \frac{107}{20} = 5\cdot35,$$

$$\overline{Y} = \frac{104}{20} = 5\cdot20.$$

By formula (5B)

$$\sigma_x = \sqrt{\left(\frac{673}{20} - 5\cdot35^2\right)} = 2\cdot242,$$

$$\sigma_y = \sqrt{\left(\frac{660}{20} - 5\cdot20^2\right)} = 2\cdot441,$$

$$\frac{S(XY)}{N} = \frac{595}{20} = 29\cdot75.$$

Hence, by formula (16A),

$$r = \frac{29\cdot75 - 5\cdot35 \times 5\cdot20}{2\cdot242 \times 2\cdot441}$$

$$= 0\cdot353.$$

6.v. If the actual values of X and Y are large, a good deal of arithmetic would be needed to obtain the third, fourth and fifth columns in the above method. Provided the ranges of the X's and Y's are fairly small, a modification of the method may be made by writing down the values of X and Y as they deviate from suitable arbitrary origins. Such arbitrary origins would be chosen near the middle of the range of each variable and the deviations of each X and Y from these would have to be written with due regard to sign. For example, if the values of X varied from 61 to 78 we might take 70 as arbitrary origin. In this case a reading of 61 would be written as -9, i.e. $61-70$. In the same way, 68 would become -2, 78 would be 8 and 70 would be 0.

In this manner we could replace our original X and Y columns by two other columns recording the deviations of the X's and Y's from their arbitrary origins. Call these X' and Y'. From these the columns of squares and products may be obtained as before. In employing this method it is advantageous to set out the plus and minus values of X' and Y' in separate columns, as shown in Example 18.

In this case $S(X')/N$ would give the difference between the arbitrary origin and the true mean of X, so that using the notation of Section 2.iii we may write $S(X')/N = D_x$, and $S(Y')/N = D_y$. Similarly, from Section 3.viii, with modifications, we get

$$\sigma_x = \sqrt{\left(\frac{S(X'^2)}{N} - D_x^2\right)} \quad \text{and} \quad \sigma_y = \sqrt{\left(\frac{S(Y'^2)}{N} - D_y^2\right)}.$$

Formula (16 A) is accordingly modified to the following in this case:

$$r = \frac{\dfrac{S(X'Y')}{N} - D_x D_y}{\sigma_x \sigma_y}. \tag{16B}$$

The calculation of r by this method is exemplified below.

Example 18. Find the correlation between the two test scores given below for 10 subjects.

Subject	1	2	3	4	5	6	7	8	9	10
Test *A*	19	25	17	20	26	30	29	21	23	24
Test *B*	145	151	140	144	138	140	142	150	149	150

Test A has a range from 17 to 30, so that a convenient arbitrary origin will be 25. For test B a convenient origin will be 145, as the range is from 138 to 150.

X'		Y'		X'^2	Y'^2	$X'Y'$	
+	−	+	−			+	−
	−6	0		36	0	0	
0		6		0	36	0	
	−8		−5	64	25	40	
	−5		−1	25	1	5	
1			−7	1	49		−7
5			−5	25	25		−25
4			−3	16	9		−12
	−4	5		16	25		−20
	−2	4		4	16		−8
	−1	5		1	25		−5
10	−26	20	−21	188	211	45	−77
−16		−1				−32	

$$D_x = -16/10 = -1\cdot6; \quad D_y = -1/10 = -0\cdot1,$$

$$\sigma_x = \sqrt{(188/10 - 1\cdot6^2)} = 4\cdot030,$$

$$\sigma_y = \sqrt{(211/10 - 0\cdot1^2)} = 4\cdot592.$$

Hence $$r = \frac{-32/10 - (-1\cdot6 \times -0\cdot1)}{4\cdot030 \times 4\cdot592} = -0\cdot183.$$

6.vi. Product-moment correlation when N is large. If N is large, over 80, the foregoing methods of calculating r become very laborious and it is usual to curtail the arithmetic involved by the use of a tabular method. For this purpose we construct what is known as a *correlation table*. The construction of such a table is most easily understood by reference to an example such as is given in Table IV, which illustrates the correlation between two tests, X and Y.

In this table each of the test scores has been grouped for working purposes (see Section 2.iii). For test X the convenient group unit was 2 and accordingly the X groupings are written along the top of the table. Test Y had a larger range and the group unit chosen was 10: the Y groupings are written at the left-hand side of the table.

We next proceed to make a spot diagram showing the distribution of the pairs of scores. For example, one subject scored 0 in test X and 55 in test Y; accordingly a dot is put in the square, or *cell*, with the X grouping 0, 1 above it and the Y grouping 50–59 on the left of it. Similarly a dot is made in the appropriate cell for each other pair of scores. Hence each dot represents a score both in test X and in test Y and the total number of dots will be N, the number of subjects who took the tests. The number of dots in each cell is counted and the number written in the cell. (If the data are on cards, the cards may be sorted into their appropriate groups and counted, thus obviating the necessity of making a spot diagram first.)

The total number of observations in each horizontal row, or *array*, is then found and recorded on the right of the table. These will form a column headed f_y, which will give the grouped frequency distribution of Y. Similarly, at the foot of each column of the table the total number of observations is recorded, giving a horizontal row for f_x, or the grouped frequency distribution of X. The totals of this last column and row, i.e. $S(f_y)$ and $S(f_x)$, should be the same and equal to N.

We then choose an arbitrary origin for each test, call the corresponding group 0, and then number the groups on both sides as we did in Section 2.iii. This will give us a column on the right headed 'y' and a row at the top of the table which will be 'x'. The means and standard deviations of both X and Y *in working units* may now be calculated, using the method of Section 3.viii. This is most conveniently done at the side of the table (see Example 19). There is no need to obtain the true means and standard deviations of the tests—indeed it is important to keep all the calculations in working units throughout. Hence we calculate D_x and D_y, using the notation of Section 2.iii, and $\sigma_{x\omega}$ and $\sigma_{y\omega}$, using the notation of formula (7 A).

There now remains the problem of finding the sum of the xy products. Reference to Table IV will show how this is done.

Starting with the top horizontal array, write in each cell the product of the frequency in that cell and the x-grouping above it. For instance, the frequency in the first cell is 1 (written in the bottom right-hand corner) and the x-grouping for that cell is -6; hence we write -6 in the top right-hand corner of the cell. Continue this for each cell containing observations. Then add these small top right-hand corner numbers for each horizontal array and record the totals in a column on the right of the table. Head this column $T_{x.y}$, indicating the total of the x's for each y. As a check on the arithmetic so far, sum this

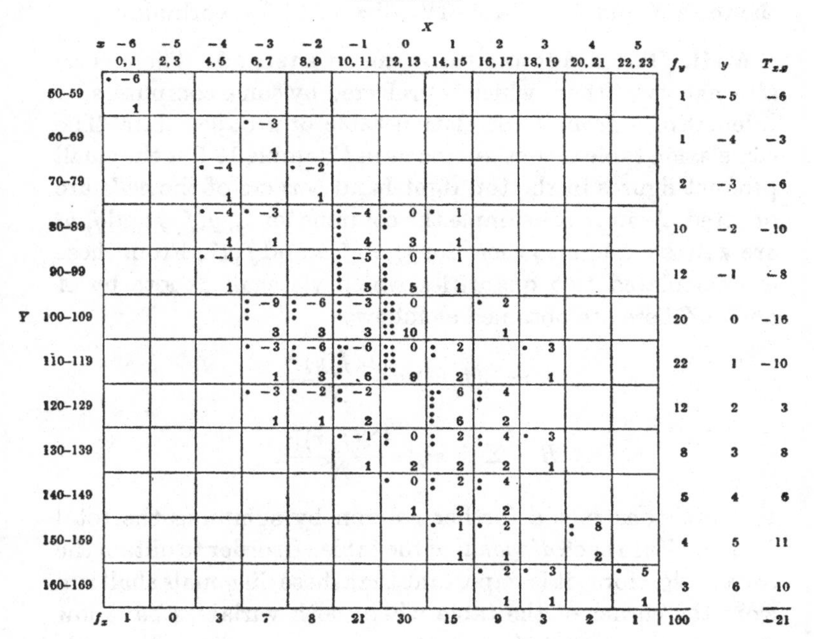

TABLE IV

Y \ X	x −6	−5	−4	−3	−2	−1	0	1	2	3	4	5	f_y	y	$T_{x.y}$
	0, 1	2, 3	4, 5	6, 7	8, 9	10, 11	12, 13	14, 15	16, 17	18, 19	20, 21	22, 23			
50–59	−6 / 1												1	−5	−6
60–69				−3 / 1									1	−4	−3
70–79			−4 / 1		−2 / 1								2	−3	−6
80–89			−4 / 1	−3 / 1		−4 / 4	0 / 3	1 / 1					10	−2	−10
90–99			−4 / 1			−5 / 5	0 / 5	1 / 1					12	−1	−8
100–109				−9 / 3	−6 / 3	−3 / 3	0 / 10		2 / 1				20	0	−16
110–119				−3 / 1	−6 / 3	−6 / 6	0 / 9	2 / 2		3 / 1			22	1	−10
120–129				−3 / 1	−2 / 1	−2 / 2		6 / 6	4 / 2				12	2	3
130–139						−1 / 1	0 / 2	2 / 2	4 / 2	3 / 1			8	3	8
140–149							0 / 1	2 / 2	4 / 2				5	4	6
150–159								1 / 1	2 / 1		8 / 2		4	5	11
160–169									2 / 1	3 / 1		5 / 1	3	6	10
f_x	1	0	3	7	8	21	30	15	9	3	2	1	100		−21

column, and the total of it, $\Sigma(T_{x.y})$, should be equal to $\Sigma(f_x x)$, as was found in calculating the mean of x.

Finally, we make one further column on the right of the table by multiplying together the entries in the y and $T_{x.y}$ columns: this new column will be headed $yT_{x.y}$. Sum this column to obtain $\Sigma(yT_{x.y})$. This gives us the sum of the xy products. Care should be taken with the signs in all these calculations.

We have now all the necessary data and the correlation coefficient may be found by substitution in the modified formula:

$$r = \frac{\frac{\Sigma(yT_{x.y})}{N} - D_x D_y}{\sigma_{x\omega} \cdot \sigma_{y\omega}}. \qquad (16\,\text{C})$$

The whole of the arithmetic involved is shown in the following example.

Example 19. Calculate the coefficient of correlation between the tests X and Y in Table IV. (See p. 59 for working.)

6.vii. The diagonal summation method. There is an alternative method, which is preferred by some computers, of calculating r from a correlation table of grouped data. The correlation table is constructed as in Example 19 but the small product figures in the top right-hand corners of the cells are omitted. As in that example the columns for f_y, y, $f_y y$ and $f_y y^2$ are written, and also those for f_x, x, $f_x x$ and $f_x x^2$. From these are calculated two quantities which we shall denote by A and B. These are obtained as follows:

$$A = \Sigma(f_y y^2) - \frac{[\Sigma(f_y y)]^2}{N},$$

$$B = \Sigma(f_x x^2) - \frac{[\Sigma(f_x x)]^2}{N}.$$

We now construct a further column by summing the total frequencies in each *diagonal* of the table. In order to obtain the correct sign for r, it is important that these diagonals shall run from the corner of the table where both variables have low scores to the corner where both have high scores; in Example 19, for instance, the diagonals will all be parallel to that running from the top left-hand corner to the bottom right-hand corner. This column we head f_d.

We then proceed as though we were finding the standard deviation of this column, i.e. an arbitrary origin is chosen and numbered 0 and the frequencies numbered positively and negatively on the two sides of this origin, yielding a column headed d. Two further columns, headed $f_d d$ and $f_d d^2$, are

Test X (each cell: frequency, with the value of x for that cell in brackets)

Test Y	x: −6 0, 1	−5 2, 3	−4 4, 5	−3 6, 7	−2 8, 9	−1 10, 11	0 12, 13	1 14, 15	2 16, 17	3 18, 19	4 20, 21	5 22, 23	f_y	y	$T_{x.y}$	$yT_{x.y}$	$f_y y$	$f_y y^2$	f_x	x	$f_x x$	$f_x x^2$
50–59	1 (−6)												1	−5	−6	30	−5	25	1	−6	−6	36
60–69				1 (−3)									1	−4	−3	12	−4	16	0	−5	0	0
70–79			1 (−4)		1 (−2)								2	−3	−6	18	−6	18	3	−4	−12	48
80–89			1 (−4)	1 (−3)		4 (−4)	3 (0)	1 (1)					10	−2	−10	20	−20	40	7	−3	−21	63
90–99			1 (−4)			5 (−5)	5 (0)	1 (1)					12	−1	−8	8	−12	12	8	−2	−16	32
100–109				3 (−9)	3 (−6)	3 (−3)	10 (0)		1 (2)				20	0	−16	0	−47	0	21	−1	−21	21
110–119				1 (−3)	3 (−6)	6 (−6)	9 (0)	2 (2)		1 (3)			22	1	−10	−10	22	22	30	0	−76	0
120–129				1 (−3)	1 (−2)	2 (−2)		6 (6)	2 (4)				12	2	3	6	24	48	15	1	15	15
130–139						1 (−1)	2 (0)	2 (2)	2 (4)	1 (3)			8	3	8	24	24	72	9	2	18	36
140–149							1 (0)	2 (2)	2 (4)				5	4	6	24	20	80	3	3	9	27
150–159								1 (1)	1 (2)		2 (8)		4	5	11	55	20	100	2	4	8	32
160–169									1 (2)	1 (3)		1 (5)	3	6	10	60	18	108	1	5	5	25
f_x	1	0	3	7	8	21	30	15	9	3	2	1	100		−21	247	128	541	100		55	335
																	−47				−76	
																	81				−21	

$$D_x = \frac{-21}{100} = -0\cdot21,$$

$$D_y = \frac{81}{100} = 0\cdot81,$$

$$\sigma_{x_w} = \sqrt{\frac{335}{100} - (-0\cdot21)^2} = \sqrt{3\cdot35 - 0\cdot0441} = 1\cdot818,$$

$$\sigma_{y_w} = \sqrt{\frac{541}{100} - 0\cdot81^2} = \sqrt{5\cdot41 - 0\cdot6561} = 2\cdot181.$$

$$r = \frac{\frac{247}{100} - (-0\cdot21 \times 0\cdot81)}{1\cdot818 \times 2\cdot181} = \frac{2\cdot47 + 0\cdot1701}{3\cdot965058} = 0\cdot666.$$

obtained in the usual way and summed. From these totals we then obtain a third quantity which we shall denote by C. This is given by

$$C = \Sigma(f_d d^2) - \frac{[\Sigma(f_d d)]^2}{N}.$$

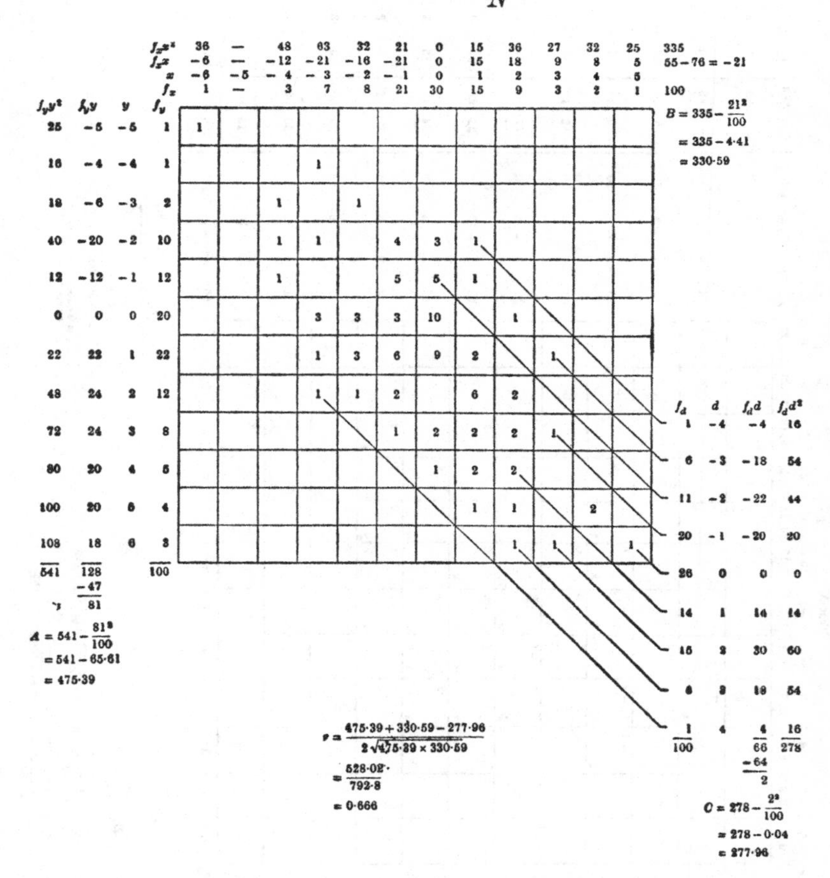

$f_y y^2$	$f_y y$	y	f_y \ x	−6	−5	−4	−3	−2	−1	0	1	2	3	4	5	
			$f_x x^2$	36	—	48	63	32	21	0	15	36	27	32	25	335
			$f_x x$	−6	—	−12	−21	−16	−21	0	15	18	9	8	5	55 − 76 = −21
			f_x	1	—	3	7	8	21	30	15	9	3	2	1	100
25	−5	−5	1	1												
16	−4	−4	1				1									
18	−6	−3	2			1		1								
40	−20	−2	10			1	1		4	3	1					
12	−12	−1	12			1			5	5	1					
0	0	0	20				3	3	3	10		1				
22	22	1	22				1	3	6	9	2		1			
48	24	2	12				1	1	2		6	2				
72	24	3	8						1	2	2	2	1			
80	20	4	5							1	2	2				
100	20	5	4								1	1		2		
108	18	6	3									1	1		1	
541	128		100													
	−47															
	81															

$B = 335 - \frac{21^2}{100}$
$= 335 - 4{\cdot}41$
$= 330{\cdot}59$

$A = 541 - \frac{81^2}{100}$
$= 541 - 65{\cdot}61$
$= 475{\cdot}39$

f_d	d	$f_d d$	$f_d d^2$
1	−4	−4	16
6	−3	−18	54
11	−2	−22	44
20	−1	−20	20
26	0	0	0
14	1	14	14
15	2	30	60
6	3	18	54
1	4	4	16
100		66	278
		−64	
		2	

$r = \frac{475{\cdot}39 + 330{\cdot}59 - 277{\cdot}96}{2\sqrt{475{\cdot}39 \times 330{\cdot}59}}$
$= \frac{528{\cdot}02}{792{\cdot}8}$
$= 0{\cdot}666$

$C = 278 - \frac{2^2}{100}$
$= 278 - 0{\cdot}04$
$= 277{\cdot}96$

The coefficient of correlation may then be obtained from the formula

$$r = \frac{A + B - C}{2\sqrt{(AB)}}. \qquad (16\text{D})$$

(It is left to the student to relate this formula to formula (16). He should have no difficulty in doing this if he bears in mind that if X and Y represent deviations of the two variables from their means, then

$$2S(XY) = S(X^2) + S(Y^2) - S(X-Y)^2;$$

and that $(X-Y)$ is constant for any diagonal.)

The whole of the arithmetic involved is shown in the following example.

Example 20. Calculate the coefficient of correlation of the data in Table IV using the method of diagonal summation.

Since in this case the f_d column will have to be written on the right of the table, it is convenient to write the f_y and associated columns on the left and the f_x and associated columns above the table. The setting out of the working is shown on p. 60.

6.viii. Significance of product-moment correlation: standard error of *r*. The standard error of r is usually calculated from the formula

$$\text{S.e. of } r = \frac{1-r^2}{\sqrt{N}}. \tag{17}$$

As usual, the probable error is 0·67449 times the standard error.

This formula is approximately true when N is large and the values of r are small or moderate in size. In such cases the correlation may be taken as differing significantly from zero if r is more than twice its standard error, or more than three times its probable error.

In small samples, however, or with very large values of r, the above formula is not true and the significance of r should be assessed by the t method. For correlation coefficients

$$t = r\frac{\sqrt{(N-2)}}{\sqrt{(1-r^2)}}. \tag{18}$$

It is unnecessary to calculate t for each value of r that is found. Appendix B gives graphically the criterion of significance for r for values of N from 50 to 270. The significance of r when N is 50 or less may be found from the following table, extracted from R. A. Fisher (Ref. 2), Table V A, or Fisher and Yates (Ref. 5), Table VI.

TABLE V. *Values of r for P* = 0·05

$n = N - 2$

n	r	n	r
1	0·997	14	0·497
2	0·950	15	0·482
3	0·878	16	0·468
4	0·811	17	0·456
5	0·755	18	0·444
6	0·707	19	0·433
7	0·666	20	0·423
8	0·632	25	0·381
9	0·602	30	0·349
10	0·576	35	0·325
11	0·553	40	0·304
12	0·532	45	0·288
13	0·514	50	0·273

In using the above table n is 2 less than N, the number of pairs of observations in the correlation. If the calculated value of r is as big as or bigger than the value given in the table for the appropriate value of n, the correlation differs significantly from zero, i.e. it indicates a real degree of association between the two variables.

The use of these various assessments of significance is illustrated in the following example.

Example 21. The scores in two tests on 100 subjects are correlated and the value of r obtained is 0·35. Is the correlation significant?

(*a*) By formula (17):

$$\text{S.e. of } r = \frac{1 - 0{\cdot}1225}{10} = 0{\cdot}08775.$$

Hence r is significant, since 0·35 is just about 4 times its standard error.

(*b*) By formula (18):

$$t = \frac{0{\cdot}35\,\sqrt{(98)}}{\sqrt{(1 - 0{\cdot}1225)}} = 3{\cdot}70.$$

This value of t is much larger than that given in Table III, p. 37, for $n = 30$, so that it must be even larger than that for $n = 98$; hence r is significantly larger than zero.

(*c*) By consulting the graph in Appendix B it will be seen that for $N = 100$ a value of r of 0·193 is significant. Hence our present value of 0·35 is definitely significant.

6.ix. Significance of the difference between two correlations. We sometimes wish to know whether the correlation between two variables is different in two different samples. To do this we make use of a method devised by Fisher (Ref. 2) which entails transforming r into a quantity which he calls z. This is given by the formula

$$z = \tfrac{1}{2}\{\log_e(1+r) - \log_e(1-r)\}.$$

Once again there is no need to make the actual calculation as Table VI gives values of z corresponding to values of r up to 0·500, and vice versa. For values outside the range of this table the student is referred to Table V B in Fisher (Ref. 2), or Table VII in Fisher and Yates (Ref. 5).

TABLE VI. *Conversion of r into z and z into r*

r	For z add	z	For r subtract
0·000–0·114	0·000	0·000–0·114	0·000
0·115–0·163	0·001	0·115–0·165	0·001
0·164–0·194	0·002	0·166–0·196	0·002
0·195–0·216	0·003	0·197–0·220	0·003
0·217–0·235	0·004	0·221–0·240	0·004
0·236–0·251	0·005	0·241–0·256	0·005
0·252–0·265	0·006	0·257–0·271	0·006
0·266–0·277	0·007	0·272–0·285	0·007
0·278–0·288	0·008	0·286–0·297	0·008
0·289–0·299	0·009	0·298–0·309	0·009
0·300–0·309	0·010	0·310–0·320	0·010
0·310–0·318	0·011	0·321–0·330	0·011
0·319–0·327	0·012	0·331–0·339	0·012
0·328–0·335	0·013	0·340–0·348	0·013
0·336–0·343	0·014	0·349–0·357	0·014
0·344–0·350	0·015	0·358–0·365	0·015
0·351–0·357	0·016	0·366–0·373	0·016
0·358–0·364	0·017	0·374–0·381	0·017
0·365–0·371	0·018	0·382–0·389	0·018
0·372–0·377	0·019	0·390–0·396	0·019
0·378–0·383	0·020	0·397–0·403	0·020
0·384–0·388	0·021	0·404–0·409	0·021

TABLE VI (*continued*)

r	For z add	z	For r subtract
0·389–0·393	0·022	0·410–0·416	0·022
0·394–0·399	0·023	0·417–0·422	0·023
0·400–0·404	0·024	0·423–0·428	0·024
0·405–0·409	0·025	0·429–0·434	0·025
0·410–0·414	0·026	0·435–0·440	0·026
0·415–0·419	0·027	0·441–0·446	0·027
0·420–0·423	0·028	0·447–0·452	0·028
0·424–0·428	0·029	0·453–0·457	0·029
0·429–0·432	0·030	0·458–0·463	0·030
0·433–0·436	0·031	0·464–0·468	0·031
0·437–0·441	0·032	0·469–0·473	0·032
0·442–0·445	0·033	0·474–0·478	0·033
0·446–0·449	0·034	0·479–0·483	0·034
0·450–0·453	0·035	0·484–0·488	0·035
0·454–0·456	0·036	0·489–0·493	0·036
0·457–0·460	0·037	0·494–0·498	0·037
0·461–0·464	0·038	0·499–0·502	0·038
0·465–0·467	0·039	0·503–0·507	0·039
0·468–0·471	0·040	0·508–0·512	0·040
0·472–0·474	0·041	0·513–0·516	0·041
0·475–0·478	0·042	0·517–0·520	0·042
0·479–0·481	0·043	0·521–0·525	0·043
0·482–0·484	0·044	0·526–0·529	0·044
0·485–0·488	0·045	0·530–0·533	0·045
0·489–0·491	0·046	0·534–0·537	0·046
0·492–0·494	0·047	0·538–0·542	0·047
0·495–0·497	0·048	0·543–0·546	0·048
0·498–0·500	0·049	0·547–0·550	0·049

To use this table, look up the value of r in the left-hand column and *add* to it the corresponding value in the second column, as z is always bigger than r. To turn z into r, look up z in the third column and *subtract* the corresponding entry in the last column.

For a value of N greater than 3, the standard error of z is $1/\sqrt{(N-3)}$, and the standard error of the difference between two z's is

$$\sqrt{\left(\frac{1}{N_1-3}+\frac{1}{N_2-3}\right)},$$

where N_1 and N_2 are the numbers of pairs in the two samples. As usual, if the difference between the two z's is greater than twice its standard error, then the difference is significant.

Example 22. Two groups of children, one of average age 11 and the other of average age 14, are given an intelligence test and an arithmetic test, and the scores are correlated for each group separately. The numbers in the groups and the correlation coefficients were as follows:

$$\begin{array}{lll} \text{11 year olds,} & N_1 = 43, & r_1 = 0{\cdot}48, \\ \text{14 year olds,} & N_2 = 39, & r_2 = 0{\cdot}39. \end{array}$$

Can the correlation between intelligence and arithmetic be regarded as different in the two groups?

From Table VI,

$$z_1 = 0{\cdot}523 \quad \text{and} \quad z_2 = 0{\cdot}412.$$

The difference is 0·111.

$$\text{S.e. of difference} = \sqrt{\left(\frac{1}{40}+\frac{1}{36}\right)} = 0{\cdot}230.$$

Hence the difference between the z's is just less than half its standard error and so cannot be regarded as significant.

6.x. Mean of several values of *r*. Use may also be made of the z transformation to obtain the average of several values of r. Coefficients of correlation should not be regarded as ordinary numbers which may be added and divided, and due weight should be given to the number of pairs in each correlation, values calculated from larger samples being more important than values from smaller samples.

To average several values of r, first transform each r into z and then multiply each z by $N-3$, where N is the number of pairs in the original r. Sum these products and divide the total by the sum of the $(N-3)$'s, giving the mean value of z. Finally transform this mean z back into r, and this will be the correct value for the mean of the original correlation coefficients. The calculation may be conveniently tabulated as in the following example.

Example 23. The same two variables are correlated in three different groups. The numbers in the groups and the values of

r are given below. What is the average correlation in the three groups?

	N	r
Group 1	23	0·41
Group 2	28	0·35
Group 3	35	0·50

Tabulate the work as follows:

r	z	$N-3$	$(N-3)z$
0·41	0·436	20	8·720
0·35	0·365	25	9·125
0·50	0·549	32	17·568
		77	35·413

$$\text{Mean } z = \frac{35{\cdot}413}{77} = 0{\cdot}460.$$

For $z = 0{\cdot}460$, $r = 0{\cdot}430$ (from Table VI). Hence the average correlation in the three groups is 0·43.

The significance of an average r may be tested as though it had been calculated from $S(N-3)+3$ pairs of observations. In the above example the average r may be tested for significance as though it had been a single r calculated from 80 pairs. It is therefore definitely significant, although of its component r's only that in group 3 is significant by itself.

6.xi. Partial correlation. In some cases it seems probable that two variables X and Y are correlated partly on account of the fact that each of them is correlated with a third variable, Z. For instance, it may be that there is a correlation between scores in an arithmetic examination and scores in a Latin examination, partly because ability to do both arithmetic and Latin is correlated with intelligence. In such a case we may wish to find the correlation between X and Y quite apart from the influence of Z. This may readily be done by the method of *partial correlation*. All we need to know is the three correlations, viz. r_{XY}, between X and Y, r_{XZ} between X and Z and r_{YZ} between Y and Z. The correlation between X and Y with the influence of Z removed is then given by the formula

$$r_{XY.Z} = \frac{r_{XY} - r_{XZ}.r_{YZ}}{\sqrt{(1-r_{XZ}^2)}\sqrt{(1-r_{YZ}^2)}}. \tag{19}$$

The symbol $r_{XY.Z}$ is read 'the correlation between X and Y keeping Z constant'.

A table giving the values of $(1-r^2)$ for all values of r to 3 places of decimals may be found in *Tables for Statisticians and Biometricians*, Table VIII, p. 20 (Ref. 3).

Example 24. Three tests, A, B and C, were given to a group of students and the three sets of scores were correlated with each other, giving the following coefficients:

$$r_{AB} = 0{\cdot}66; \quad r_{AC} = 0{\cdot}60; \quad r_{BC} = 0{\cdot}40.$$

What is the correlation between A and B keeping C constant?

From formula (19),

$$r_{AB.C} = \frac{0{\cdot}66 - 0{\cdot}60 \times 0{\cdot}40}{\sqrt{(1-0{\cdot}36)}\sqrt{(1-0{\cdot}16)}}$$

$$= \frac{0{\cdot}42}{0{\cdot}7328}$$

$$= 0{\cdot}57.$$

The above formula and method may be extended to four or more variables. The formula for four variables is given below: the student is unlikely to need further extensions.

$$r_{AB.CD} = \frac{r_{AB.C} - r_{AD.C} \cdot r_{BD.C}}{\sqrt{(1-r^2_{AD.C})}\sqrt{(1-r^2_{BD.C})}}. \tag{19 A}$$

It will be seen that this partial correlation, the correlation between A and B keeping both C and D constant, requires the calculation of three other partial correlations, each keeping only one variable constant.

6.xii. Significance of a partial correlation coefficient. The significance of a partial correlation coefficient may be determined by calculating t, which in this case is given by

$$t = \frac{r_p}{\sqrt{(1-r_p^2)}}\sqrt{(N-p-2)}, \tag{20}$$

where r_p is the partial coefficient and p is the number of variables held constant. The number of degrees of freedom for consulting the table of t (Table III) is $n = N-p-2$. Thus in Example 24, if there were 39 students in the group, we have $r_p = 0{\cdot}57$, $p = 1$, and $N = 39$.

Hence
$$t = \frac{0\cdot57}{\sqrt{(1-0\cdot57^2)}}\sqrt{(39-1-2)}$$
$$= \frac{0\cdot57\times6\cdot0}{\sqrt{0\cdot6751}}$$
$$= 4\cdot17.$$

Reference to Table III with $n = 39-1-2 = 36$ shows the coefficient to be clearly significant.

EXERCISES ON CHAPTER VI

20. Calculate the coefficient of correlation by the product-moment method between:

(*a*) subjects 1–25 in tests C and D,
(*b*) subjects 26–50 in tests C and D,
(*c*) subjects 51–75 in tests C and D,
(*d*) subjects 76–100 in tests C and D.

Use formula (16A).

21. Check the results of Exercise 20 by using the method of Section 6.v and formula (16B). In each case take 25 as arbitrary origin for test C and 30 as arbitrary origin for test D.

22. Calculate the coefficient of correlation between each test from A to G inclusive and each other one for the whole 100 subjects. Employ the method of the correlation table, Section 6.vi, using formula (16C) and the grouping adopted in Exercise 3. Repeat using the method of diagonal summation and formula (16D).

23. Using formula (17), calculate the standard error of the values of r obtained in Exercise 20 for the correlations between test F and the other tests from A to G. Hence determine the significance of these values of r and check by reference to (*a*) Table V and (*b*) the graph in Appendix B.

24. Using the correlation coefficients given in the answer to Exercise 20, determine whether r_{BE} is significantly different from r_{CE}, r_{GE}, r_{EF}, r_{FG}, r_{CF}, and r_{DF}. Use the method of Section 6.ix and Table VI.

25. From the correlation coefficients given in the answer to Exercise 20, calculate the following partial correlation coefficients:

(*a*) between D and F, keeping C constant;
(*b*) between C and F, keeping D constant;
(*c*) between F and G, keeping E constant;
(*d*) between E and F, keeping G constant;
(*e*) between C and E, keeping B constant.

Chapter VII

OTHER METHODS OF CORRELATION

7.1. The ranking method. (a) Spearman's coefficient. It sometimes happens that the actual values of two variables cannot be accurately measured, although we are able to rank them in order of size or merit. In such a case the method of product-moment correlation cannot be applied, but an approximate coefficient of correlation may be calculated. If N pairs of variables are ranked, X and Y being ranked separately of course, and d represents the difference between the ranks of X and Y for any one pair, then the *coefficient of ranked correlation* is given by the formula

$$\rho = 1 - \frac{6\Sigma(d^2)}{N(N^2-1)}. \qquad (21)$$

This formula may be used whether the distributions of X and Y are normal or not, but it only gives an approximate indication of the association between the two variables and should never be used if it is possible to calculate r. The coefficient should not be employed in partial correlation, multiple correlation, factorial analysis or any other statistical process which is based on product-moment correlation.

The method of ranked correlation is frequently employed because the calculation involved is simple for small samples. The method of calculation is given here for use in cases where the data are inadequate for product-moment correlation.

The first step is to rank each variable, calling the best or biggest value 1, the second best or biggest 2, and so on. When two or more values of a variable are the same it is usual to give each the average rank. For instance, if there are two equal values for the 6th place, each is ranked as $6\frac{1}{2}$, the next rank being 8, since these two will have occupied the 6th and 7th places between them. Similarly, if there are three equal values for the 10th place, each will be ranked as 11, since they will occupy the 10th, 11th and 12th places between them, and the next value will be ranked as 13. By this means, if there are N pairs of variables, each variable will be ranked from 1 to N. (This averaging of ranks introduces a further inaccuracy into formula (21), for the formula assumes that each ranking is different: it is, however, frequently impossible to avoid it in using this method.)

The pairs of ranks are written down in two columns and a third column, headed d, formed by subtracting entries in the second column from corresponding entries in the first. If the correct signs are put down in this column, the total of the column, or $\Sigma(d)$, should be zero. Each entry in the third column is then squared, yielding a fourth column headed d^2. Finally, this column is summed to obtain $\Sigma(d^2)$, which is then substituted in formula (21).

In Appendix C are given values of the reciprocal of $6/N(N^2-1)$ for values of N from 10 to 60. The use of this, as explained in the Appendix, will save the student a good deal of arithmetic.

When N is moderately large, the significance of ρ may be tested by calculating t, which is given in this case by

$$t = \rho\sqrt{\frac{N-2}{1-\rho^2}}. \qquad (21\text{A})$$

In referring to Table III, $n = N-2$.

Example 25. A class of 15 schoolboys was given an intelligence test and the master provided their order of merit in an entrance examination. What is the correlation between the test and the entrance examination? Below are the marks and order of merit of each boy.

Boy	*A*	*B*	*C*	*D*	*E*	*F*	*G*	*H*
Order of merit ...	1	2	3	4	5	6	7	8
Intelligence test score	22	19	6	18	20	16	11	9

Boy	*I*	*J*	*K*	*L*	*M*	*N*	*O*
Order of merit ...	9	10	11	12	13	14	15
Intelligence test score	15	12	10	7	13	12	8

Here one of the variables is ranked for us and may be written straight down. As regards the marks for the test, boy A scores most and is given the rank 1; *E* is next and is ranked 2; *B* is 3, and so on, finally giving us the second column.

Boy	Order of merit rank	Test rank	d	d^2
A	1	1	0	0
B	2	3	− 1	1
C	3	15	−12	144
D	4	4	0	0
E	5	2	3	9
F	6	5	1	1
G	7	10	− 3	9
H	8	12	− 4	16
I	9	6	3	9
J	10	8½	1½	2¼
K	11	11	0	0
L	12	14	− 2	4
M	13	7	6	36
N	14	8½	5½	30¼
O	15	13	2	4
			22	265½
			−22	
			0	

$$\rho = 1 - \frac{6 \times 265 \cdot 5}{15 \times 224} = 1 - 0 \cdot 474 = 0 \cdot 526.$$

Examining the significance of this value of ρ we have by formula (21 A),

$$t = 0 \cdot 526 \sqrt{\frac{13}{0 \cdot 723324}} = 2 \cdot 23.$$

This is just greater than the critical value of 2·13 given in Table III for $n = 15$, hence we may conclude that ρ is significant.

(b) Kendall's coefficient. Suppose we have n objects ranked for each of two variables, X and Y. For convenience these rankings may be written horizontally as below:

	A	*B*	*C*	*D*	*E*	*F*	*G*	*H*
X	1	2	3	4	5	6	7	8
Y	1	5	11	2	3½	8	6	3½

	I	*J*	*K*	*L*	*M*	*N*	*O*
X	9	10	11	12	13	14	15
Y	13	7	11	9	15	14	11

There are 15 objects ranked here for X and Y; and for simplicity, though this is not necessary, the X rankings are arranged in ascending order.

The essence of Kendall's method, which yields a coefficient of ranked correlation called τ (tau), is to compare each pair of ranks with each other pair and to allot to each a score of 1, 0 or -1. Let X_i and X_j be the ith and jth rankings for X, where X_j is to the right of X_i. If X_i is smaller than X_j a score of 1 is allotted to that pair: if they are equal, the score is 0, and if X_i is the larger, the score is -1. Call the score for this pair a_{ij}. Repeat the process for the Y rankings and call the score for the Y_iY_j pair b_{ij}. With n objects there will be a total of $\binom{n}{2}$ pairs. For each pair, multiply a_{ij} and b_{ij} together and sum the products over the whole $\binom{n}{2}$ pairs to obtain $\Sigma(a_{ij}b_{ij})$. Then τ, the Kendall coefficient of ranked correlation, is given by the formula

$$\tau = \frac{\Sigma(a_{ij}b_{ij})}{\frac{1}{2}n(n-1)}. \tag{22}$$

This is the value of the coefficient *when there are no tied rankings*. When there are ties, as in the Y rankings above, the denominator of formula (22) needs to be modified. There may be several sets of tied ranks in one row, perhaps 2 ranks being tied in one set, 3 in another, and so on. If t is the number of ranks tied in any one set, calculate $\frac{1}{2}t(t-1)$ for that set. Repeat for all sets of ties in that row and add all the results. Call the total U_a for the X row and U_b for the Y row. Then the coefficient, when there are ties, is given by

$$\tau = \frac{\Sigma(a_{ij}b_{ij})}{\sqrt{\{[\frac{1}{2}n(n-1)-U_a]\,[\frac{1}{2}n(n-1)-U_b]\}}}. \tag{22 A}$$

Example 26. Calculate τ for the data at the beginning of this section.

First compare the ranking for A with that for B. In the X row, 1 is less than 2 so that $a_{ij} = 1$. In the Y row, 1 is less than 5, so that b_{ij} is also 1. Hence for the pair AB,

$$a_{ij}b_{ij} = 1 \times 1 = 1.$$

Next compare A with C, A with D and so on, until A has been compared with each of the others B to O. Add the 14 values of $a_{ij}b_{ij}$ thus obtained and record the total. We have now finished with the A rankings. Next compare the B rankings with each of the 13 others, C to O, in the same way. Some of these will be found to be negative, e.g. for BD, $a_{ij} = 1$ and $b_{ij} = -1$, so that $a_{ij}b_{ij} = -1$. Also, for the pair EH, for example, $a_{ij} = 1$ and $b_{ij} = 0$, so that $a_{ij}b_{ij} = 0$. Listing the various sub-totals we find the following:

Compared with remainder	Sub-total ($a_{ij}b_{ij}$)
A	14
B	7
C	-4
D	11
E	9
F	3
G	6
H	7
I	-2
J	5
K	1
L	3
M	-2
N	-1

Add the various sub-totals which gives us $\Sigma(a_{ij}b_{ij})$, which in this case is equal to 57.

There are no ties in the X row, so that $U_a = 0$. In the Y row there are two sets of ties. E and H are tied: for these $t = 2$ and $\frac{1}{2}t(t-1) = 1$. C, K and O are also tied: for these $t = 3$ and $\frac{1}{2}t(t-1) = 3$. Hence $U_b = 1+3 = 4$. Substituting in formula (22 A) we find

$$\tau = \frac{57}{\sqrt{\{105(105-4)\}}} = 0{\cdot}554.$$

(For the same data, as the student may verify, $\rho = 0{\cdot}717$. It will be found that as a rule the numerical value of τ is less than that of ρ.)

In the above example the X ranking was written out in order. This is not essential, indeed actual ranking is not necessary; for instance, below are the scores of 7 subjects in two tests, X and Y:

	A	B	C	D	E	F	G
X	45	51	52	47	60	48	61
Y	13	15	12	10	16	13	18

Pairs of scores may be compared and values for a_{ij} and b_{ij} allotted as usual, dependent simply upon whether the second score in each pair is larger, equal to or smaller than the first. For the above data the student should find that $\tau = 0{\cdot}586$. If now a further subject, H, were tested and obtained scores of 57 in X and 17 in Y, τ for all 8 subjects could be calculated by comparing the scores of subjects A to G with those for H and adding the sub-total for $a_{ij}b_{ij}$ thus obtained to the $\Sigma(a_{ij}b_{ij})$ for the 7 subjects, and remembering that n in the denominator of formula (22 A) is now 8 instead of 7. The student may verify that for all 8 subjects, $\tau = 0{\cdot}618$.

This illustrates a very useful property of Kendall's method of calculating the coefficient of ranked correlation, viz. that additional data may be added and τ calculated without having to re-rank each time, as is necessary in Spearman's method. If our data were in a time series, therefore, a fresh pair of readings being made each week, say, we could calculate a running value for τ and see how it varied with time.

7.ii. Significance of τ. To test the significance of τ we first need to calculate the variance of $\Sigma(a_{ij}b_{ij})$, which we will call var Σ. When there are no tied rankings this is given by

$$\text{var}\,\Sigma = \frac{n(n-1)(2n+5)}{18}.^* \qquad (23)$$

Take the square root of this, which will give the standard deviation of Σ, and then if the ratio of $\Sigma(a_{ij}b_{ij})-1$ to $\sqrt{\text{var}\,\Sigma}$ is greater than $1{\cdot}65$, the value of τ is significant.

Example 27. Test the significance of a value of τ of $0{\cdot}667$ There were 10 pairs of readings and $\Sigma(a_{ij}b_{ij})$ was 30.

* See Appendix F for a table to help in the calculation of var Σ.

From formula (23),

$$\operatorname{var}\Sigma = 10\times 9\times 25/18 = 125,$$
$$\sqrt{\operatorname{var}\Sigma} = 11{\cdot}18$$
$$\frac{\Sigma(a_{ij}b_{ij})-1}{\sqrt{\operatorname{var}\Sigma}} = \frac{30-1}{11{\cdot}18} = 2{\cdot}59.$$

This is greater than 1·65 so that τ is significant.

When there are tied rankings in either or both rows, the expression for $\operatorname{var}\Sigma$ is more complicated. Suppose there are ties to the extent of t_1, t_2, etc., in the X row; for each set calculate $t(t-1)(2t+5)$ and sum the sets to obtain

$$\Sigma t(t-1)(2t+5).$$

Similarly for the Y row, calling the ties u_1, u_2, etc., we obtain $\Sigma u(u-1)(2u+5)$. Other adjustments involving the values of t and u have to be made, so that the complete expression for $\operatorname{var}\Sigma$ when there are ties present is

$$\begin{aligned}\operatorname{var}\Sigma = \tfrac{1}{18}\{n(n-1)(2n+5) - \Sigma t(t-1)(2t+5) \\ - \Sigma u(u-1)(2u+5)\} \\ + \frac{1}{9n(n-1)(n-2)}\{\Sigma t(t-1)(t-2)\}\{\Sigma u(u-1)(u-2)\} \\ + \frac{1}{2n(n-1)}\{\Sigma t(t-1)\}\{\Sigma u(u-1)\}.^{*}\end{aligned} \qquad (23\,\text{A})$$

Using this the significance of τ may be tested as before.

7.iii. Partial rank correlation. Using the Kendall coefficients of ranked correlation it is possible to calculate partial correlation coefficients. It happens that the formula is similar to formula (19) for partial product-moment correlation, in fact

$$\tau_{AB.C} = \frac{\tau_{AB} - \tau_{AC}\cdot\tau_{BC}}{\sqrt{[\{1-\tau_{AC}^2\}\{1-\tau_{BC}^2\}]}}.$$

As yet no tests for the significance of a partial τ are known. No similar expression using Spearman coefficients is available.

7.iv. Biserial correlation. Observational data sometimes make it impossible to calculate either product-moment or ranked correlation coefficients. For example, a test may be applied to a group of subjects about whom the only other in-

* See Appendices F and G for tables to help in the calculation.

formation we possess is that each of them has either passed or failed a particular examination. In such a case, after making two assumptions, we may calculate a form of correlation coefficient known as the *coefficient of biserial correlation*. This we shall denote *bis. r*. The two necessary assumptions are:

(1) that the dichotomous variable is normally distributed, and

(2) that the regression of X on Y is linear (see Chapter VIII). If both these assumptions are deemed justifiable and if we have more than 80 observations, so that the data may be grouped, the *bis. r* coefficient may be calculated as follows.

Suppose X is the numerical variable, i.e. the test, and Y the dichotomous variable, i.e. the variable divided into only two parts. Choose a convenient group unit for X and divide the observations into the appropriate groups. Then construct a two-row, or biserial, table showing how the subjects who pass or fail in Y fall into the X-groups. Such a table might appear as under:

Test X

		−5	−4	−3	−2	−1	0	1	2	3	4	5	6	7	Total
Y	Pass	0	1	0	4	5	9	9	10	6	3	4	2	2	55
	Fail	1	0	2	6	7	9	11	4	2	2	0	1	0	45
	Total	1	1	2	10	12	18	20	14	8	5	4	3	2	100

As in calculating the standard deviation, we choose an arbitrary origin for X and number off the groups on both sides, as in Section 2.iii. This has been done in the table above.

Now let us call the portion of the x's which fall into the *larger* part of the Y distribution, x_1. Then the mean of this row (which will be the Passes in the above table) will be $\bar{x}_1$. The mean of all the x's will be $\bar{x}$ and their standard deviation σ_x. All these statistics may be calculated, *in working units*, from the table. The corresponding statistics for Y cannot be calculated directly, but assuming that we knew them, the coefficient of biserial correlation would be given by the formula

$$bis.\, r = \frac{\dfrac{\bar{x}_1 - \bar{x}}{\sigma_x}}{\dfrac{\bar{y}_1 - \bar{y}}{\sigma_y}}. \tag{24}$$

Although we cannot calculate $\frac{\bar{y}_1-\bar{y}}{\sigma_y}$ directly from the table, we may obtain it indirectly (on the above assumptions) from data given in Table II of the *Tables for Statisticians and Biometricians* (Ref. 3). This table gives a quantity $\frac{1}{2}(1+\alpha)$* and the values of a function z corresponding to them. (This z is not to be confused with the z used by Fisher as a transformation for r, as in Section 6.ix.) The quantity $\frac{1}{2}(1+\alpha)$ in our case is equal to n_1/N, where n_1 is the number of observations in the larger portion of Y and N is the total number of observations in the whole table. This is readily calculated and the value of z corresponding to it may be found by interpolation. (The method of doing this may be best understood from Example 28.)

We then have
$$\frac{\bar{y}_1-\bar{y}}{\sigma_y}=\frac{z}{n_1/N}.$$

The coefficient of biserial correlation is therefore obtained by calculating $\frac{\bar{x}_1-\bar{x}}{\sigma_x}$ from the two-row table and dividing this by the value of $\frac{\bar{y}_1-\bar{y}}{\sigma_y}$ obtained from the statistical tables.

The sign of *bis.* r has to be determined by inspection. If, for instance, the mean x score of the Passes is greater than that of the Failures, then the correlation is positive.

This method of correlation should not be resorted to if it is possible to avoid it. In most cases it gives little more information than would be acquired by investigating the significance of the x-score means difference of the two Y-groups. In any case, the ordinary standard error of r does not apply to *bis.* r, and caution is needed in interpreting the coefficient. (The standard error of *bis.* r is known only to a first approximation and the student who wishes to look further into this matter is referred to H. E. Soper, *Biometrika*, vol. x, p. 384, 1914.)

The method of calculation is shown in the example below.

* If a vertical is drawn at any point x in a normal curve, the total area is divided into two unequal portions, if x deviates from the mean. $\frac{1}{2}(1+\alpha)$ is the area of the greater portion.

Example 28. Calculate the coefficient of biserial correlation for the two-row table given above.

The total column at the foot of the table gives the grouped frequency distribution of x, so that $\bar{x}$ and σ_x may be calculated by the method of Section 3.viii.

f	x	fx	fx^2
1	−5	−5	25
1	−4	−4	16
2	−3	−6	18
10	−2	−20	40
12	−1	−12	12
		−47	
18	0		0
20	1	20	20
14	2	28	56
8	3	24	72
5	4	20	80
4	5	20	100
3	6	18	108
2	7	14	98
100		144	645
		−47	
		97	

$$\bar{x}(=D) = \frac{97}{100} = 0\cdot97,$$

$$\sigma_x = \sqrt{\left(\frac{645}{100} - 0\cdot97^2\right)}$$

$$= 2\cdot347.$$

The mean $\bar{x}_1$ is obtained in a similar manner by multiplying x by the entries in the top row: these we may call f_1.

f_1	x	f_1x
0	−5	0
1	−4	−4
0	−3	0
4	−2	−8
5	−1	−5
		−17
9	0	
9	1	9
10	2	20
6	3	18
3	4	12
4	5	20
2	6	12
2	7	14
55		105
		−17
		88

$$\bar{x}_1 = \frac{\Sigma(f_1 x)}{n_1} = \frac{88}{55} = 1\cdot6.$$

From the biserial table, therefore,

$$\frac{\bar{x}_1-\bar{x}}{\sigma_x} = \frac{1{\cdot}6-0{\cdot}97}{2{\cdot}347} = 0{\cdot}2684.$$

Now we need to calculate $\frac{\bar{y}_1-\bar{y}}{\sigma_y}$ from the tables in Ref. 3. $n_1 = 55$ and $N = 100$. Hence

$$\tfrac{1}{2}(1+\alpha) = 55/100 = 0{\cdot}55.$$

From the table we find the following values of z corresponding to two values of $\frac{1}{2}(1+\alpha)$:

$$\text{for}\quad \tfrac{1}{2}(1+\alpha) = 0{\cdot}5477584,\quad z = 0{\cdot}3960802,$$
$$\text{for}\quad \tfrac{1}{2}(1+\alpha) = 0{\cdot}5517168,\quad z = 0{\cdot}3955854.$$

Our value of $\frac{1}{2}(1+\alpha)$ lies between these two. Now the difference between the two values of $\frac{1}{2}(1+\alpha)$ given above is 0·0039584. This corresponds to a difference in z of 0·0004948.

Our value of $\frac{1}{2}(1+\alpha)$ is $0{\cdot}55-0{\cdot}5477584 = 0{\cdot}0022416$ above the smaller of the two values given above. It will be noticed that as $\frac{1}{2}(1+\alpha)$ gets larger, z gets smaller: hence we have to *subtract* from the first z that part of the difference between the two z's which is proportional to 0·0022416/0·0039584, i.e. the value of z corresponding to $\frac{1}{2}(1+\alpha) = 0{\cdot}55$ is

$$\begin{aligned} &0{\cdot}3960802-0{\cdot}0004948\times 0{\cdot}0022416/0{\cdot}0039584 \\ &= 0{\cdot}3960802-0{\cdot}0002802 \\ &= 0{\cdot}3958000.^* \end{aligned}$$

Hence $$\frac{\bar{y}_1-\bar{y}}{\sigma_y} = \frac{z}{n_1/N} = \frac{0{\cdot}3958}{0{\cdot}55} = 0{\cdot}7196.$$

Therefore $$bis.\,r = \frac{\dfrac{\bar{x}_1-\bar{x}}{\sigma_x}}{\dfrac{\bar{y}_1-\bar{y}}{\sigma_y}} = \frac{0{\cdot}2684}{0{\cdot}7196} = 0{\cdot}373.$$

This coefficient must be positive since $\bar{x}_1$, the mean test score of the Passes, is larger than $\bar{x}$, the mean of all the subjects, and so must be even larger than the mean of the Failures alone.

* This method of linear interpolation is not strictly applicable to the table but is sufficiently approximate for the present purpose.

Hence the Passes have a better score than the Failures and the relationship between the test scores and the examinational success is a positive one.

7.v. Fourfold correlation. When both variables are dichotomous the methods previously described cannot be applied. There are various methods of calculating fourfold or tetrachoric correlation coefficients, but their use is not advised. The association between two dichotomous variables is best investigated by the method described in Section 9.iii, p. 99.

EXERCISES ON CHAPTER VII

26. (*a*) Using the Spearman ranking method (formula (21)), calculate the coefficient of ranked correlation between the order of merit H and tests C and D for the subjects 1–25 in Appendix E. (Rank the *smallest* order of merit as 1 and the *largest* test score as 1.)

(*b*) From the rankings of the subjects obtained above, calculate the Spearman coefficient of ranked correlation between tests C and D for subjects 1–25. Compare this with the value of r obtained in Exercise 20 (*a*).

27. Calculate the Kendall coefficient of ranked correlation for the data of Example 25, p. 70.

28. Calculate the Kendall coefficients of ranked correlation for the same data as in Exercise 26 above.

Chapter VIII

REGRESSION AND THE CORRELATION RATIO

8.i. Graphic construction of regression lines. When we are considering the correlation between two variables, X and Y, we may draw a graph showing the mean values of y for regularly increasing values of x. This graph will be an irregular line which is called the *observed regression line* of y on x. Similarly the observed regression line of x on y is an irregular line showing the mean values of x for regularly increasing values of y. These lines indicate the law of change in the mean of one variable for unit change in the other, and if the lines are straight the regressions are said to be *linear*.

Usually, owing to errors of sampling, the observed regression lines are rather irregular, but it may be possible to 'fit' straight lines to them and to show mathematically that the observed regression lines do not depart significantly from the fitted straight regression lines.

A straight line has an algebraic equation which represents it in symbols; hence linear regression lines may be represented by the following equations:

$$(y-\bar{y}) = r\frac{\sigma_y}{\sigma_x}(x-\bar{x}), \tag{25}$$

$$(x-\bar{x}) = r\frac{\sigma_x}{\sigma_y}(y-\bar{y}). \tag{25 A}$$

The former is the regression straight line of y on x and the latter the regression straight line of x on y. In these equations r is the coefficient of correlation between X and Y. In (25) x is called the *independent* and y the *dependent* variable, and vice versa in (25 A).

The angles these lines make to the horizontal and vertical respectively are measured by the expressions

$$r\frac{\sigma_y}{\sigma_x} \quad \text{and} \quad r\frac{\sigma_x}{\sigma_y},$$

and these are called the *coefficients of regression*.

If $r = 1$, the two regression equations are identical and the regression straight lines coincide. If $r = 0$, the two lines are horizontal and vertical respectively and cross at right angles. The lines cross for any intermediate value of r, so that the larger the value of r, the smaller is the acute angle between them. The two lines always cross at the point $\bar{x}$, $\bar{y}$ on the graph, i.e. the point indicating where the means of x and y lie.

In any fairly small sample the successive observed means of one variable for different values of the other are unlikely to fall exactly on a straight line, but if the regression does not depart significantly from linearity it is possible to draw a straight line which passes very nearly through the observed means. As an example, the regression straight lines of the data in Example 19, p. 58, will be drawn and the discrepancies between the observed regression lines and these may be observed.

At the right-hand side of the correlation table in that example we find two columns headed f_y and $T_{x.y}$ respectively. If we divide the entries in the second of these by the corresponding entries in the first, we obtain the mean values of x corresponding to each y-group. In Example 19 these are as follows:

y	f_y	$T_{x.y}$	$\frac{T_{x.y}}{f_y}$
−5	1	−6	−6·00
−4	1	−3	−3·00
−3	2	−6	−3·00
−2	10	−10	−1·00
−1	12	−8	−0·67
0	20	−16	−0·80
1	22	−10	−0·45
2	12	3	0·25
3	8	8	1·00
4	5	6	1·20
5	4	11	2·75
6	3	10	3·33

The last column gives the observed line of regression of x on y. In like manner the observed line of regression of y on x may be obtained by calculating a $T_{y.x}$ column from the correlation table and dividing the entries in it by the corresponding entries in the f_x column. This gives us for the observed regression of y on x:

x	f_x	$T_{y.x}$	$\frac{T_{y.x}}{f_x}$
−6	1	−5	−5·00
−5	0	—	—*
−4	3	−6	−2·00
−3	7	−3	−0·43
−2	8	2	0·25
−1	21	0	0·00
0	30	8	0·27
1	15	30	2·00
2	9	29	3·22
3	3	10	3·33
4	2	10	5·00
5	1	6	6·00

* Note that there are no observations in the $x = -5$ column. There will therefore be no entry in the last column for this group. Care must be taken *not* to record the entry as 0·00, as this would be taken as a point on the observed regression line.

We now proceed to plot these points on a graph. Since we are working from arbitrary origins in this example, the two axes will be at right angles, crossing at the point $x = 0$, $y = 0$. The x line will be horizontal and we mark off along it equal divisions corresponding to the x-groups. Those to the right of the origin, or point where the axes cross, will be positive and those to the left negative, and they are numbered accordingly. Similarly in the case of y, divisions above the origin are positive and those below negative.

In Fig. 4 the points on the observed regression lines are plotted, those for the regression of x on y being marked by crosses and those for the regression of y on x by circles. The regression straight lines themselves, *AA* and *BB*, are also drawn. It will be seen that the crosses mostly lie closely about the line *AA* and the circles about the line *BB*. These two lines cross at the point where $x = -0{\cdot}21$ and $y = 0{\cdot}81$, which were the mean values of x and y *in working units* found in Example 19. The acute angle between the lines indicates a fairly high correlation between x and y; it was found in fact that $r = 0{\cdot}666$.

The lines AA and BB are drawn as follows. The regression coefficients are first calculated from the data $r = 0{\cdot}666$, $\sigma_x = 1{\cdot}818$ and $\sigma_y = 2{\cdot}181$.

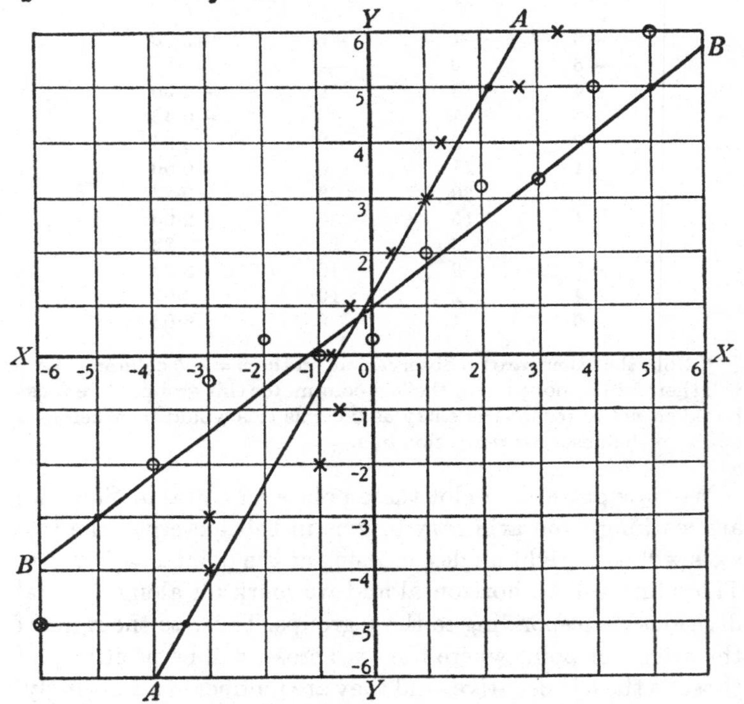

Fig. 4. Regression lines of data in Example 19.

$$\text{Hence} \qquad r\frac{\sigma_x}{\sigma_y} = 0{\cdot}666 \times 1{\cdot}818/2{\cdot}181 = 0{\cdot}5551$$

$$\text{and} \qquad r\frac{\sigma_y}{\sigma_x} = 0{\cdot}666 \times 2{\cdot}181/1{\cdot}818 = 0{\cdot}7990.$$

Substituting these in the regression equations, (25) and (25 A), we get

$$(y - \bar{y}) = 0{\cdot}7990(x - \bar{x}),$$

and

$$(x - \bar{x}) = 0{\cdot}5551(y - \bar{y}).$$

Now we know that

$$\bar{x} = -0{\cdot}21 \quad \text{and} \quad \bar{y} = 0{\cdot}81.$$

Hence we have

$$y-0{\cdot}81 = 0{\cdot}7990(x+0{\cdot}21)$$

and $$x+0{\cdot}21 = 0{\cdot}5551(y-0{\cdot}81),$$

whence $$y = 0{\cdot}7990x+0{\cdot}9778$$

and $$x = 0{\cdot}5551y-0{\cdot}6596.$$

By substituting different values of x and y in these equations and calculating the corresponding values of y and x, the fitted regression lines may be plotted exactly. Two points are enough to give each line.

In the first equation, for example,

$$\text{if } x = -5, \quad y = -3{\cdot}0172$$

and $$\text{if } x = 5, \quad y = 4{\cdot}9728.$$

These two points fix the line BB.

Similarly the line AA is obtained from the second equation:

$$\text{if } y = -5, \quad x = -3{\cdot}4351$$

and $$\text{if } y = 5, \quad x = 2{\cdot}1159.$$

8.ii. Testing the linearity of regressions. Since the method of product-moment correlation is usually appropriate only in those cases where the regressions of the two variables on each other are linear, it is important to be able to ascertain whether in fact the regressions are linear in any particular case. Often the graphic method exemplified above suffices, but in cases of doubt (and always more satisfactorily) the linearity of regression may be tested mathematically. The process involves what is called the 'Analysis of Variance'. (See Chapter XI.)

Essentially the method is as follows. The student will remember that the variance of x is obtained from the sum of the squares of the deviations of each individual x from the general mean, $\bar{x}$. This sum of squares may be split into two portions:

(1) the sum of the squares of the deviations of each x from the mean of its own array, summed for all arrays, and

(2) the sum of the squares of the deviations of the means of the arrays from the general mean, $\bar{x}$. In the latter each array mean must be weighted by the number of observations in it.

This splitting of the variance may be shown symbolically. Let f_x be the number of observations in a y-array corresponding to a particular value of x, and $\bar{y}_x$ be the mean of the y's in this array. It may then be shown that

$$S(y-\bar{y})^2 = S(y-\bar{y}_x)^2 + \Sigma[f_x(\bar{y}_x-\bar{y})^2],$$

where Σ means a summation for all different values of x.

Now if there is a linear regression of y on x, there will be a mean y for each array which may be calculated from the regression equation. The quantity $\Sigma[f_x(\bar{y}_x-\bar{y})^2]$ may therefore be subdivided into two portions, one of which is the sum of squares due to linear regression and the other the weighted sum of the squares of the deviations of the array means from the means calculated on the assumption of linear regression. If this latter portion of the variance is sufficiently large, it signifies that the means of the arrays differ significantly from linear regression, and the following section shows how the significance of this departure from linearity may be tested arithmetically.

8.iii. In Example 19, ten columns were constructed on the right of the correlation table. In order to examine the linearity of the regression of y on x for the same data, we need to construct three further columns. The first of these is the $T_{y.x}$ column, as shown in Section 8.i. The second is obtained by dividing entries in this first column by corresponding entries in the f_x column: this is headed $T_{y.x}/f_x$. Entries in this second column are then multiplied by corresponding entries in the first to give a final column headed $(T_{y\ x})^2/f_x$.

From the sums of certain of these thirteen columns we have to calculate three quantities:

$$(1)\quad \Sigma[f_x(\bar{y}_x-\bar{y})^2],$$

$$(2)\quad S(y-\bar{y})^2,$$

$$(3)\quad \frac{[S(x-\bar{x})(y-\bar{y})]^2}{S(x-\bar{x})^2}.$$

These can be calculated as follows:

$$(1)\quad \Sigma[f_x(\bar{y}_x-\bar{y})^2] = \Sigma\left\{\frac{(T_{y.x})^2}{f_x}\right\} - \frac{(\Sigma T_{y.x})^2}{N},$$

(2) $$S(y-\bar{y})^2 = \Sigma(f_y y^2) - \frac{(\Sigma T_{y.x})^2}{N},$$

(3) $$\frac{[S(x-\bar{x})(y-\bar{y})]^2}{S(x-\bar{x})^2} = \frac{\left[\Sigma(yT_{x.y}) - \frac{\Sigma(f_x x).\Sigma(f_y y)}{N}\right]^2}{\Sigma(f_x x^2) - \frac{\{\Sigma(f_x x)\}^2}{N}}.$$

From these data we construct a table. In computing the number of degrees of freedom in each part of the variance, a is the number of y-arrays.

	Degrees of freedom	Mean square
A. Total sum of squares: $S(y-\bar{y})^2$	$N-1$	
B. Total sum of squares within arrays: $(A-C)$	$N-a$	$\frac{A-C}{N-a}$
C. Total sum of squares between arrays: $\Sigma[f_x(\bar{y}_x-\bar{y})^2]$	$a-1$	
D. Sum of squares due to deviations from linear regression: $(C-E)$	$a-2$	$\frac{C-E}{a-2}$
E. Sum of squares due to linear regression: $\frac{[S(x-\bar{x})(y-\bar{y})]^2}{S(x-\bar{x})^2}$	1	

In this table entries B and D are obtained by subtraction as indicated, B by subtracting C from A, and D by subtracting E from C. The entries in B and D are then divided by their respective degrees of freedom, giving the two entries in the column headed 'Mean square'.

*We next find the logarithms of the entries in the mean square column and multiply the *difference* between these logarithms by 1·1513 (to convert logarithms to the base 10 to Napierian logarithms and to divide by 2). This gives us a quantity called z, which is tabulated in Table VI of Fisher's book. (Unfortunately this is the third z we have had to use and should not be confused with those of Sections 6.ix and 7.iv.)

Hence

$$z = 1{\cdot}1513\left|\log\frac{A-C}{N-a} - \log\frac{C-E}{a-2}\right|.$$

* See next section.

It finally remains to look up the value for z in Fisher's table corresponding to the number of degrees of freedom we have. There are two values of n required, n_1 and n_2. The n_1 of the table is the number of degrees of freedom corresponding to the *larger* mean square, i.e. whichever of the entries B or D is the bigger, and n_2 is the number of degrees of freedom corresponding to the smaller mean square.

If the calculated z is *smaller* than the z in Fisher's 5 % point table, then the regression of y on x does not differ significantly from linearity.

The whole of the necessary calculation is shown in Example 29. It must be remembered that this is an examination of the regression of y on x only: the regression of x on y should also be examined by an exactly similar method, interchanging x and y in each formula.

Example 29. Examine the linearity of the regression of y on x for the correlation table given in Example 19.

For the sake of space the correlation table itself is not reproduced here. The first ten columns to the right of the table are copied from Example 19, p. 58, and were obtained as explained in the section preceding that example.

From the calculations on p. 89 we may construct the analysis table. The number of arrays, a, = 12.

		Degrees of freedom	Mean square
A. Total sum of squares	475·39	99	
B. Total sum of squares within arrays	227·3033	88	2·5830
C. Total sum of squares between arrays	248·0867	11	
D. Sum of squares due to deviations from linear regression	37·2467	10	3·7247
E. Sum of squares due to linear regression	210·84	1	

$$z = 1{\cdot}1513\,(\log 3{\cdot}7247 - \log 2{\cdot}5830) = 0{\cdot}1830.$$

We must now consult Fisher's 5 % point table of z. We have $n_1 = 10$ and $n_2 = 88$.

Now we see from the table that when $n_1 = 12$ and $n_2 = 120$, the 5 % point is 0·3032, hence the value for $n_1 = 10$ and $n_2 = 88$

f_y	y	$T_{x.y}$	$yT_{x.y}$	$f_y y$	$f_y y^2$	f_x	x	$f_x x$	$f_x x^2$	$T_{y.x}$	$\frac{T_{y.x}}{f_x}$	$\frac{(T_{y.x})^2}{f_x}$
1	−5	−6	30	−5	25	1	−6	−6	36	−5	−5·0	25·0
1	−4	−3	12	−4	16	0	−5	0	0	—	—	—
2	−3	−6	18	−6	18	3	−4	−12	48	−6	−2·0	12·0
10	−2	−10	20	−20	40	7	−3	−21	63	−3	−0·4286	1·2857
12	−1	−8	8	−12	12	8	−2	−16	32	2	0·25	0·5
20	0	−16	0	−47	0	21	−1	−21	21	0	0	0
22	1	−10	−10	22	22	30	0	−76	0	8	0·2667	2·1333
12	2	3	6	24	48	15	1	15	15	30	2·0	60·0
8	3	8	24	24	72	9	2	18	36	29	3·2222	93·4444
5	4	6	24	20	80	3	3	9	27	10	3·3333	33·3333
4	5	11	55	20	100	2	4	8	32	10	5·0	50·0
3	6	10	60	18	108	1	5	5	25	6	6·0	36·0
100		−21	247	128	541	100		55	335	81		313·6967
				−47				−76				
				81				−21				

$$S(y-\bar{y})^2 = 541-\frac{81^2}{100} = 541-65{\cdot}61 = 475{\cdot}39,$$

$$\frac{[S(x-\bar{x})(y-\bar{y})]^2}{S(x-\bar{x})^2} = \frac{\left(247+\frac{21\times 81}{100}\right)^2}{335-\frac{21^2}{100}} = \frac{264{\cdot}01^2}{330{\cdot}59} = 210{\cdot}84,$$

$$\Sigma[f_x(\bar{y}_x-\bar{y})^2] = 313{\cdot}6967-\frac{81^2}{100} = 248{\cdot}0867.$$

must be bigger than this. Accordingly our calculated value of z is much smaller than the 5 % point and we conclude that the regression of y on x does not depart significantly from linearity.

8.iv. The method of calculating z at the end of the previous section was included for the sake of theoretical completeness. In actual practice the rather cumbrous calculation may be avoided by the use of a table given in Fisher and Yates's tables (Ref. 5). Instead of working out the value of z we may calculate the *variance ratio*. In testing the linearity of regression by this method, what we wish to show is whether or not the variance due to deviations from linearity of regression is significantly *greater* than that within arrays. If the former variance is *smaller* than the latter, there is no need to make a test at all—there is no evidence of departure from linearity of regression. (The student need not concern himself with the very rare case where the variance due to deviations from linear regression is significantly smaller than the variance within arrays.)

The variance ratio in this case is the entry in the 'Mean square' column in the D row divided by that in the B row. If this ratio is greater than 1 we must consult Table V for the 5 % point in the Fisher and Yates tables. For this purpose n_1 in the table is the number of degrees of freedom corresponding to the *larger* variance, i.e. that in the D row. If the calculated ratio is *smaller* than the appropriate entry in the table, then the regression does not depart significantly from linearity.

Taking the data of Example 29, for instance, the variance ratio is $3{\cdot}7247/2{\cdot}5830 = 1{\cdot}44$. In this case $n_1 = 10$ and $n_2 = 88$. In Table V of Fisher and Yates for $n_1 = 12$ and $n_2 = 120$ we find an entry of 1·83 for the 5 % point. Our calculated variance ratio is definitely smaller than this and so would be relatively even smaller than the one corresponding to the number of degrees of freedom in our data. (The correct 5 % point ratio could be obtained from the table by interpolation, but there is no need for this.) We conclude, therefore, that the regression does not depart significantly from linearity, which agrees with our findings in the previous section.

8.v. The correlation ratio. If the regressions of the two variables on each other are non-linear, the degree of association between the variables cannot be measured by ordinary correlation. There might be a real relationship and yet the coefficient of correlation would be zero if the regressions were semicircular and symmetrical, for example. In such cases a measure of association may be obtained by calculating the *correlation ratio.*

There are two correlation ratios for each pair of variables and they are denoted by η_{yx} and η_{xy}. The formulae for these are simple:

$$\eta_{yx}^2 = \frac{\sigma_{\bar{y}_x}^2}{\sigma_y^2} = \frac{\Sigma[f_x(\bar{y}_x - \bar{y})^2]}{S(y-\bar{y})^2}, \tag{26}$$

$$\eta_{xy}^2 = \frac{\sigma_{\bar{x}_y}^2}{\sigma_x^2} = \frac{\Sigma[f_y(\bar{x}_y - \bar{x})^2]}{S(x-\bar{x})^2}. \tag{26 A}$$

In these formulae $\sigma_{\bar{x}_y}$ signifies the standard deviation of the means of the x-arrays and $\sigma_{\bar{y}_x}$ the standard deviation of the y-arrays, so that η may be seen to be the ratio between the standard deviation of the means of arrays and the standard deviation of the whole sample.

It will be seen that the requisite data for the calculation of η_{yx} have already been obtained in Example 29, and if the student has examined the linearity of the regression of x on y he will also have the data for calculating η_{xy}.

Example 30. Calculate η_{yx} from the data obtained in Example 29.

From that example we see that

$$\Sigma[f_x(\bar{y}_x - \bar{y})^2] = 248{\cdot}0867 \quad \text{and} \quad S(y-\bar{y})^2 = 475{\cdot}39.$$

Hence
$$\eta_{yx}^2 = \frac{248{\cdot}0867}{475{\cdot}39} = 0{\cdot}52186,$$

whence
$$\eta_{yx} = \sqrt{0{\cdot}52186} = 0{\cdot}7224.$$

8.vi. The data necessary for the examination of the linearity of regression may now be expressed in an alternative form. We have

$$S(y-\bar{y})^2 = N\sigma_y^2,$$

$$\Sigma[f_x(\bar{y}_x - \bar{y})^2] = N\eta_{yx}^2\sigma_y^2,$$

$$\frac{[S(x-\bar{x})(y-\bar{y})]^2}{S(x-\bar{x})^2} = Nr^2\sigma_y^2.$$

Also the z of Section 8.iii may be expressed as

$$z = 1{\cdot}1513 \log \frac{(N-a)(\eta^2 - r^2)}{(a-2)(1-\eta^2)}.$$

If we substitute in this expression the values we have obtained, we find that

$$z = 1{\cdot}1513 \log \frac{88(0{\cdot}7224^2 - 0{\cdot}666^2)}{10(1-0{\cdot}7224^2)}$$

$$= 1{\cdot}1513 \log \frac{6{\cdot}89075}{4{\cdot}7814}$$

$$= 1{\cdot}1513 \log (1{\cdot}4412)$$

$$= 0{\cdot}183.$$

This result is identical with the value of z calculated in Example 29.

It may also be observed that when η and r are equal in the above expression, z is zero, i.e. when the regressions are absolutely linear, both η's are equal and equal to r.

EXERCISE ON CHAPTER VIII

29. (*a*) From the table used in Exercise 22, examine the linearity of the regressions of tests F and G on each other.

(*b*) Thence calculate the two correlation ratios for tests F and G.

Chapter IX

χ^2: CONTINGENCY: GOODNESS OF FIT

9.i. The statistical methods already described have mostly been applicable to quantitative numerical data only, and usually only to data which are at any rate approximately normally distributed. It may happen, however, that the available data are qualitative or quantitative only in the sense that we know the number of cases falling into different categories. In such instances the methods which have been explained in the previous chapters cannot be used but there are other methods which are appropriate.

These methods depend chiefly on a statistic known as χ^2. The mathematical derivation of this statistic is difficult and cannot be described here. However, the distribution of it has been worked out and tables are available showing the frequency with which different values of χ^2 are exceeded and also the values of χ^2 corresponding to particular frequencies. Reference to the appropriate tables will be made later in this chapter.

Use may be made of χ^2 in the investigation of a number of different problems, but the calculation of it is essentially the same in each case. If O is the observed frequency in a particular category into which a variable may fall and E is the frequency which would be expected to fall in that category on some hypothesis, then χ^2 may be found by dividing the square of the difference between O and E by E and summing these quotients for all categories into which the variable falls. In symbols,

$$\chi^2 = \Sigma\left[\frac{(O-E)^2}{E}\right]. \qquad (27)$$

This is the general formula and the calculation of it in specific cases will be decribed in due course.

Having found χ^2 we need to know the number of degrees of freedom available for calculating it in each particular case before we can make use of the χ^2 tables. Rules for finding the number of degrees of freedom will be given in each instance. Consultation of the tables will then indicate the probability, P, of a calculated value of χ^2 being exceeded as a result of random sampling. If this probability is less than 0·05 (i.e. 19 : 1), then χ^2 may be regarded as showing that the observed data depart significantly from the hypothesis which is being examined. This hypothesis may be that a variable has a particular type of distribution or, more frequently, that there is no association between two variables. This latter hypothesis, known as a *null hypothesis*, assumes that two variables are not associated: if we can disprove a null hypothesis, it follows that the two variables must be associated. Such a procedure is usual in the investigation of certain problems of association which will now be described.

9.ii. Contingency. Suppose x and y are two variables which are not measurable numerically but which can each be divided into two or more categories. An example of this might be hair colour and eye colour, for instance. Let the different categories of x be x_1, x_2, x_3, etc., and those of y be y_1, y_2, y_3, etc. We may then construct a *contingency* table showing the number of x_1's which fall into the y_1, y_2, y_3, etc. categories, and so on. This will give us a sort of small correlation table. An example is shown below for 5 x-categories and 4 y-categories.

	x_1	x_2	x_3	x_4	x_5	
y_1						n_{y_1}
y_2						n_{y_2}
y_3						n_{y_3}
y_4						n_{y_4}
	n_{x_1}	n_{x_2}	n_{x_3}	n_{x_4}	n_{x_5}	N

It will be seen that the table is divided into $5 \times 4 = 20$ rectangles or *cells*. The total frequency in each x-category is given at the foot of the columns and will be n_{x_1}, n_{x_2}, etc. Similarly the total frequency in each y-category is given at the end of the horizontal rows and will be n_{y_1}, n_{y_2}, etc. The total frequency of each variable, or pairs of variables, will as usual be N. In general, if there are r rows and c columns, there will be $r \times c$ cells. Each cell may be referred to by the x and y categories into which it falls: e.g. the cell on the third row down and in the second column from the left may be referred to as the x_2, y_3 cell.

The contingency table is completed by filling in the observed frequency in each cell. For instance, we count how many observations in the x_1 category are also in the y_1 category and write down the number in the x_1, y_1 cell, and so on. From the completed table we may calculate a coefficient known as the *coefficient of mean square contingency* and denoted by C. To do this we have first to calculate the frequency which would be expected in each cell on the null hypothesis, i.e. on the assumption that the two variables are not associated with one another. This is done quite simply as follows: if n_{x_c} is the total frequency in the x_c column, and n_{y_r} the total frequency in the y_r row, then the frequency that would be expected in the x_c, y_r cell on the null hypothesis is

$$\frac{n_{x_c} \times n_{y_r}}{N}.$$

By substituting each of the values of c and r in turn, we obtain the expected frequency in every cell.

The next step is to subtract the expected frequency from the observed frequency in each cell: the resulting values of $(O-E)$ are written in each cell with due regard to sign, and the arithmetic may be checked at this stage by observing that the total of $(O-E)$ is zero for each row and each column.

Next square $(O-E)$ and divide the square by E for each cell. This gives a series of quotients which is conveniently written down on the right of the table: there will be $r \times c$ such quotients. Finally this column of figures is added, giving χ^2.

From this the coefficient of contingency, C, is given by the formula

$$C = \sqrt{\frac{\chi^2}{N+\chi^2}}. \tag{28}$$

[Sometimes an intermediate step is inserted in the definition of C. χ^2 is called the 'square contingency' and if we divide it by N we get the 'mean square contingency', which is denoted by ϕ^2. Then

$$C = \sqrt{\frac{\phi^2}{1+\phi^2}}. \qquad \text{(28 A)]}$$

If the null hypothesis is correct, O and E will be equal for each cell (apart from errors of sampling), so that χ^2 will be zero and C will also be zero. It is evident from formula (28) that C can never quite equal unity. The actual maximum value of C depends on the number of cells, so that for a 2×2 table, for example, the maximum possible value of C is 0·707, and for a 10×10 table the maximum for C is 0·949. It is obvious, therefore, that a value of C obtained from one contingency table cannot be directly compared with that from another, unless the number of rows and columns in the one is equal to that in the other. Hence in reporting a value of C the number of cells in the table should always be mentioned. Moreover, C is always positive, so that the nature of the association has to be determined by inspection of the table. It may be seen, therefore, that C itself is not a very useful coefficient. Of much more use is the information to be derived from χ^2.

The standard error of C is exceedingly complex and can only be interpreted for very large samples. Accordingly it is seldom used and instead we find from the χ^2 table the corresponding value of P, which gives the probability of our calculated value of χ^2 being exceeded as the result of random sampling. There are two usual methods of doing this, making use of a table given by Fisher or one due to Elderton. Fisher's method is probably the more convenient and will be described first.

(1) First we need to know n, the number of degrees of freedom available for calculating χ^2. In the case of a contingency table of r rows and c columns this is given by $n = (r-1)(c-1)$. We may then consult Fisher's table, of which an extract is given below, noting the value of χ^2 appearing against this value of n in the column headed $P = 0·05$. If the calculated value of χ^2 is *greater* than that given in the table, then it is significant, and the null hypothesis is disproved, i.e. there is a significant association between the variables.

An extract from Fisher's table (Ref. 2, Table III) or Fisher and Yates's (Ref. 5, Table IV) for $P = 0{\cdot}05$ is given below.

TABLE VII. *Table of χ^2 for different values of n*

$P = 0{\cdot}05$

n	χ^2	n	χ^2	n	χ^2
1	3·841	11	19·675	21	32·671
2	5·991	12	21·026	22	33·924
3	7·815	13	22·362	23	35·172
4	9·488	14	23·685	24	36·415
5	11·070	15	24·996	25	37·652
6	12·592	16	26·296	26	38·885
7	14·067	17	27·587	27	40·113
8	15·507	18	28·869	28	41·337
9	16·919	19	30·144	29	42·557
10	18·307	20	31·410	30	43·773

If n is greater than 30, significance can be tested by calculating $\sqrt{(2\chi^2)} - \sqrt{(2n-1)}$. If this exceeds 1·65, χ^2 is significant.

(2) To make use of Elderton's table, which is given in *Tables for Statisticians and Biometricians* (Ref. 3, Table XII), we need to know n', which is one more than the number of degrees of freedom, i.e. $n' = (r-1)(c-1)+1$. Different values of n' are given at the heads of columns in Elderton's table and integral values of χ^2 are listed at the side. By looking up the value of P in the appropriate n' column opposite the calculated value of χ^2, we obtain the probability of as great a value of χ^2 or greater occurring as a result of random sampling; if this value of P is *less* than 0·05, χ^2 is significant. Since the listed values of χ^2 are integral, it is usually necessary to interpolate to find the value of P.

The use of both tables is illustrated in Example 31.

In using the method of contingency there are two provisions to be borne in mind. First, E, the expected frequency, must be at least 5 in each cell; secondly, the table should if possible contain at least 5 rows and 5 columns. The former provision is the more important.

Example 31. Two examiners assessed the intelligence of 200 students, one by a verbal test and the other by a performance test. Each graded the intelligence as Very Good, Good, Fair or Poor. The relation between the two sets of judgments is shown in the contingency table below. Calculate the coefficient of contingency and examine whether the relationship between the two judgments can be regarded as significant or not.

		Examiner 1			
		V.G.	G.	F.	P.
Examiner 2	V.G.	19	10	8	3
	G.	7	40	9	4
	F.	8	20	23	19
	P.	0	8	12	10

First we construct a contingency table, leaving room for three entries in each cell. Then in each cell we enter the observed frequency, the expected frequency and the difference between the two, thus:

	$O-E$
O	
	E

		V.G.	G.	F.	P.	
V.G.	$O-E$	12·2	−5·6	−2·4	−4·2	0
	O	19	10	8	3	40
	E	6·8	15·6	10·4	7·2	40
G.	$O-E$	−3·2	16·6	−6·6	−6·8	0
	O	7	40	9	4	60
	E	10·2	23·4	15·6	10·8	60
F.	$O-E$	−3·9	−7·3	4·8	6·4	0
	O	8	20	23	19	70
	E	11·9	27·3	18·2	12·6	70
P.	$O-E$	−5·1	−3·7	4·2	4·6	0
	O	0	8	12	10	30
	E	5·1	11·7	7·8	5·4	30
	$O-E$	0	0	0	0	
	O	34	78	52	36	200
	E	34	78	52	36	

$\frac{(O-E)^2}{E}$
21·888
2·010
0·554
2·450
1·004
11·776
2·792
4·281
1·278
1·952
1·266
3·251
5·100
1·170
2·262
3·918
$\chi^2 = 66·952$

$$C = \sqrt{\frac{\chi^2}{N+\chi^2}} = \sqrt{\frac{66·952}{266·952}}$$

$$= 0·501.$$

Along the bottom and at the side of the table the arithmetic is checked by showing that the total of the expected frequencies for each row and each column is the same as the total of the observed frequencies, and also that the sum of $(O-E)$ is zero for each row and each column. Then for each cell we calculate $(O-E)^2/E$ and tabulate the values obtained on the right of the table. The sum of these is χ^2. The remainder of the calculation of C is shown below the table.

We shall assess the significance of this result by both methods.

(1) Using Fisher's table:

$$n = (4-1)(4-1) = 9.$$

From Table VII, for $n = 9$, $\chi^2 = 16{\cdot}919$. The calculated value of χ^2 is much bigger than the value in the table, hence the distribution departs significantly from independence, i.e. there is a significant relation between the two sets of judgments.

(2) Using Elderton's table:

$$n' = (4-1)(4-1)+1 = 10.$$

For $n' = 10$, opposite $\chi^2 = 60$ we get a value of $P = 0{\cdot}000000$. This means that a value of χ^2 as big or bigger than the one calculated would arise as a result of random sampling less than once in a million times. Accordingly there is no doubt about the significance of the relationship between the two sets of judgments.

Note that Fisher's method is the easier for proving whether or not there is a significant relationship in a single table. If we wish to compare the significance in two or more tables we need to know the value of P for each; in this case we have to use Elderton's table or else the complete χ^2 tables given in Fisher's book.

9.iii. 2 × 2 tables. A special form of contingency arises when both variables are dichotomous, i.e. each variable can be divided into only two classes. The association between such variables may be shown by constructing a contingency table with only four cells, since there will be only two columns and two rows. We shall now examine the significance of a fourfold or 2×2 table making use of the χ^2 method.

In order to simplify the notation we shall denote the observed frequencies in the four cells by a, b, c and d as under:

	x_1	x_2	
y_1	a	b	$a+b$
y_2	c	d	$c+d$
	$a+c$	$b+d$	N

It will be seen that $N = a+b+c+d$.

The value of χ^2 may be determined in the same way as that used for any contingency table, but the arithmetic may be shortened by using the formula

$$\chi^2 = \frac{(ad-bc)^2 N}{(a+c)(b+d)(c+d)(a+b)}. \tag{29}$$

The denominator of this will be seen to be the product of the totals of the rows and columns.

Having found χ^2, its significance may be determined by consulting Fisher's table for $n = 1$. It will be seen that if the calculated value of χ^2 is as great or greater than 3·841, then there is a significant association between the two variables.

Alternatively, the value of P may be found by consulting a special table given by Yule (Ref. 1, p. 534). Reference to this table will show that for a value of χ^2 equal to 3·84, $P = 0{\cdot}05$. This, of course, agrees with Fisher's table, but if exact values of P are required for purposes of comparison they may be found from Yule's table.

As in the case of all contingency tables, the nature of the association in a 2×2 table has to be determined by inspection.

Example 32. Is there a significant association between X and Y in the following data?

		X: x_1	x_2	Total
Y	y_1	65	25	90
	y_2	20	50	70
	Total	85	75	160

From formula (29)

$$\chi^2 = \frac{(65 \times 50 - 25 \times 20)^2 \times 160}{85 \times 75 \times 70 \times 90}.$$

This may readily be evaluated by logarithms and we find that

$$\chi^2 = 30{\cdot}13.$$

This value is much greater than 3·84, the critical significance value of χ^2 for one degree of freedom, so that we may conclude that there is a significant relationship between X and Y.

9.iv. Yates's correction. The χ^2 distribution is a continuous one but in 2×2 tables the number of sets of observations which can fit the marginal totals is finite and limited, so that the actual distribution from such a table is definitely discontinuous. Allowance for this fact may be made by applying *Yates's correction for continuity*. Essentially this consists of decreasing by $\frac{1}{2}$ those cell values which are greater than expectation and increasing by $\frac{1}{2}$ those which are less than expectation. This has the effect of slightly decreasing the value of χ^2.

Applying this correction, formula (29) becomes

$$\chi^2 = \frac{\left(ad - bc - \frac{N}{2}\right)^2 N}{(a+c)(b+d)(c+d)(a+b)}. \qquad (30)$$

From the data in the previous Example the student may verify that the value of χ^2 applying Yates's correction is 28·4. With such a large value the correction has not altered significance, but with borderline cases, where χ^2 is only just significant as calculated by formula (29), the correction should always be applied since it may very easily bring the value below the significance level.

9.v. Alternative method of calculation. An alternative method of calculating χ^2 from a 2×2 table, which may be used as an arithmetical check on results obtained from formula (29), is as follows:

Let p_1 be the proportion of y_1 which is in the x_1 class;
p_2 be the proportion of y_2 which is in the x_1 class;
p be the proportion of the total population in the x_1 class; and
$q = 1 - p$.

Then $$p_1 = \frac{a}{a+b}; \quad p_2 = \frac{c}{c+d}; \quad p = \frac{a+c}{N}.$$

χ^2 is then given by the formula

$$\chi^2 = \frac{p_1 a + p_2 c - p(a+c)}{pq}. \tag{31}$$

The algebraic proof of the identity of this expression with that given in formula (29) is quite simple and is left to the student as an exercise. Note that Yates's correction cannot be applied with this method of calculation.

Applying this formula to the data of Example 32 we have:

$$p_1 = 65/90, \quad p_2 = 20/70, \quad p = 85/160 \quad \text{and} \quad q = 75/160.$$

Hence $$= \frac{\dfrac{65 \times 65}{90} + \dfrac{20 \times 20}{70} - \dfrac{85 \times 85}{160}}{\dfrac{85}{160} \times \dfrac{75}{160}}$$

$$= \frac{46{\cdot}944 + 5{\cdot}714 - 45{\cdot}156}{0{\cdot}249}$$

$$= 30{\cdot}13.$$

This result can be seen to be identical with that obtained in Example 32.

(*Note.* For a way of treating 2×2 tables by the method of rank correlation see J. W. Whitfield, *Biometrika*, vol. XXXIV, December 1947, pp. 295–6.)

9.vi. $2 \times n$ tables. Cases frequently occur where we wish to examine the association between two variables, one of which is dichotomous whilst the other is capable of division into several, say n, classes. In such cases χ^2 may be calculated by the method of Section 9.ii, but the calculation of the expected frequencies may be avoided in the following manner. Suppose the $2 \times n$ table is represented as under, y being the dichotomous variable and x the variable with several categories —four in the example. The observed frequencies in the cells are denoted in each case by the letter a with appropriate suffixes and dashes:

	x_1	x_2	x_3	x_4	
y_1	a_1	a_2	a_3	a_4	n_1
y_2	a'_1	a'_2	a'_3	a'_4	n_2
	$a_1+a'_1$	$a_2+a'_2$	$a_3+a'_3$	$a_4+a'_4$	N

If we take a and a' to represent any associated pair of observed frequencies, then we calculate for each pair

$$\frac{1}{a+a'}(an_2 - a'n_1)^2,$$

where n_1 and n_2 are the total frequencies in the y_1 and y_2 classes. This expression is evaluated for each pair of associated frequencies and the sum of all such expressions is divided by $(n_1 \times n_2)$, giving χ^2. Hence

$$\chi^2 = \frac{1}{n_1 n_2} S\left\{\frac{1}{a+a'}(an_2 - a'n_1)^2\right\}. \tag{32}$$

The number of degrees of freedom in a $2 \times n$ table is $n-1$, so that the significance of χ^2 in the above case may be determined by consulting Table VII for $n = 3$.

Example 33. Determine the significance of the association between the variables x and y in the following contingency table:

	x_1	x_2	x_3	x_4	Total
y_1	36	42	30	12	120
y_2	12	22	20	26	80
Total	48	64	50	38	200

Calculating the expression $\frac{1}{a+a'}(an_2 - a'n_1)^2$ to the nearest whole number for each pair of associated frequencies, we have

$$\frac{1}{48}(36\times 80-12\times 120)^2 = \frac{2,073,600}{48} = 43,200,$$

$$\frac{1}{64}(42\times 80-22\times 120)^2 = \frac{518,400}{64} = 8,100,$$

$$\frac{1}{50}(30\times 80-20\times 120)^2 = \frac{0}{50} = 0,$$

$$\frac{1}{38}(12\times 80-26\times 120)^2 = \frac{4,665,600}{38} = 122,779.$$

The sum of these quantities = 174,079. Therefore

$$\chi^2 = \frac{174,079}{120\times 80} = 18{\cdot}1.$$

In Table VII for $P = 0{\cdot}05$ and $n = 3$ we find a value of χ^2 of 7·815. Our calculated value is much greater than this and we may therefore conclude that there is a significant association between the two variables.

An alternative method of calculation, similar to that given for 2×2 tables, may be used as a check. If p is the proportion in the y_1 class in any vertical column, i.e. $p = a/(a+a')$, and P is the proportion in the y_1 class in the whole population, so that $P = n_1/N$ and $Q = 1-P$, then

$$\chi^2 = \frac{1}{PQ}\left\{S(ap) - n_1 P\right\} = \frac{N^2}{n_1 n_2}\left\{S\left(\frac{a^2}{a+a'}\right) - \frac{n_1^2}{N}\right\}. \quad (33)$$

To calculate this, first work out $a^2/(a+a')$ for each vertical column and then substitute in the above formula. We obtain:

$$\begin{aligned} 36^2/48 &= 27{\cdot}0 \\ 42^2/64 &= 27{\cdot}5625 \\ 30^2/50 &= 18{\cdot}0 \\ 12^2/38 &= 3{\cdot}7895 \\ S[a^2/(a+a')] &= \overline{76{\cdot}3520} \end{aligned}$$

By substitution

$$\begin{aligned} \chi^2 &= \frac{200^2}{120\times 80}\left(76{\cdot}352 - \frac{120^2}{200}\right) \\ &= 4{\cdot}1667 \times 4{\cdot}352 \\ &= 18{\cdot}13. \end{aligned}$$

9.vii. Goodness of fit. One other use of the χ^2 method may be given. Suppose we have an observed frequency distribution of a variable and wish to examine the validity of some hypothesis about that distribution. We may do this by calculating what the distribution would be on that hypothesis and examining the agreement, or *goodness of fit*, of the observed and calculated distributions.

As an example, an observed frequency distribution of accidents will be given and an examination made of the hypothesis that the accidents are distributed in a *Poisson series*. If the probability of an event occurring is very small but a long enough period of observation is taken for it to happen sometimes, so that it may happen 0, 1, 2, 3, ... times, then the distribution of its occurrence will form what is called a Poisson series, if a large enough number of independent observations is made. The frequency distribution for 0, 1, 2, 3, ... occurrences will be given by the successive terms of the series

$$Ne^{-m}\left(1, m, \frac{m^2}{2!}, \frac{m^3}{3!}, \ldots\right),$$

where N is the total number of observations and m is the mean number of occurrences.

Example 34. Examine the hypothesis that the following frequency distribution of accidents forms a Poisson series:

Number of accidents	Observed frequency
0	14
1	37
2	76
3	70
4	64
5	53
6	31
7	19
8	14
9	9
10	5
11	5
12	1
	398

In order to calculate the expected frequency on the assumption that the distribution forms a Poisson series we need to know m, the mean number of accidents. This is readily calculated by the method of formula (2), Section 2.ii. We find that $\Sigma(fX) = 1549$, whence the mean number of accidents is $1549/398 = 3{\cdot}892$. Substituting this value for m and 398 for N in the above formula for the Poisson series, we obtain the expected frequencies as below (see later for method of calculation):

Number of accidents	Expected frequency	
0	8·12	
1	31·61	
2	61·51	
3	79·80	
4	77·64	
5	60·43	
6	39·20	
7	21·80	
8	10·60	
9	4·59	7·28 (9–14)
10	1·78	
11	0·63	
12	0·20	
13	0·06	
14	0·02	
	397·99	

In order to satisfy the provision that in calculating χ^2 the expected frequencies shall not be less than 5, the last six entries are grouped together. Since the expected frequencies have been calculated to only 2 places of decimals, the total of the expected frequencies is not exactly 398, but it is so very little less that it can have no marked effect on the value of χ^2. Calling the observed and expected frequencies O and E respectively, we now construct columns headed $(O-E)$, $(O-E)^2$ and $(O-E)^2/E$. The total of the last column is χ^2. All six columns are given below.

We have used ten groups to calculate χ^2 but the number of degrees of freedom is only 8, since both N and m were fixed in the formation of the Poisson series. For $n = 8$, we find that

$\chi^2 = 15{\cdot}507$ in Table VII, p. 97. The calculated value is much greater than this, hence the Poisson series is a *bad fit* to the observed accident distribution. We find, in fact, from Elderton's table that for such a value of χ^2, P is less than 0·001, and so we may conclude that the observed distribution certainly does not form a Poisson series.

No. of accidents	O	E	$(O-E)$	$(O-E)^2$	$\frac{(O-E)^2}{E}$
0	14	8·12	5·88	34·5744	4·26
1	37	31·61	5·39	29·0521	0·92
2	76	61·51	14·49	209·9601	3·41
3	70	79·80	− 9·80	96·0400	1·20
4	64	77·64	−13·64	186·0496	2·40
5	53	60·43	− 7·43	55·2049	0·91
6	31	39·20	− 8·20	67·2400	1·72
7	19	21·80	− 2·80	7·8400	0·36
8	14	10·60	3·40	11·5600	1·09
9 and over	20	7·28	12·72	161·7984	22·23
			0·01		38·50

(*Note.* The total of the $(O-E)$ column should as usual be zero. It is not exactly so in our case since the expected frequencies were calculated to only two places of decimals.)

Method of calculating the expected frequencies

The calculation of the expected frequencies in the above example is best done by using logarithms. In the Poisson series the term Ne^{-m} is common to each successive frequency, so this is worked out first:

$$\log 398 = 2{\cdot}59988,$$
$$\log e = 0{\cdot}43429,$$
$$\log\log e = \bar{1}{\cdot}63778,$$
$$\log 3{\cdot}892 = 0{\cdot}59018;$$

hence $3{\cdot}892 \log e = \text{antilog}\,(\bar{1}{\cdot}63778 + 0{\cdot}59018)$

$$= 1{\cdot}69026;$$

therefore

$$\log 398e^{-3{\cdot}892} = 2{\cdot}59988 - 1{\cdot}69026$$
$$= 0{\cdot}90962,$$

whence $398e^{-3{\cdot}892} = 8{\cdot}12.$

The values of the successive terms may then be found by a tabular method, making use of the logarithms and logarithms of factorials given in Fisher and Yates's tables (Ref. 5, pp. 62, 76).

The method is indicated below. p is the index of the power to which m is raised in the successive terms:

$\log m^p$	$\log p!$	$\log m^p - \log p!$	$(\log m^p - \log p!) + \log Ne^{-m}$	antilog
0·59018	0·0	0·59018	1·49980	31·61
1·18036	0·30103	0·87933	1·78895	61·51
1·77054	0·77815	0·99239	1·90201	79·80
etc.	etc.	etc.	etc.	etc.

9.viii. Curve fitting. Frequently in biological experiments a series of observations or groups of observations is obtained and it is wished to examine the trend of the series. A graph of the series may indicate to the eye what shape the trend appears to have, but it is often important to determine mathematically what is the algebraic equation of the line that best describes the trend. For example, a group of subjects is repeatedly tested with the same test and we may wish to define the shape of the improvement curve obtained, or we may wish to find the shape of the growth curve of experimental animals fed daily on a certain diet. Such a process is known as *curve fitting*, and in many instances the process may be reduced mathematically to a method similar to that of fitting a straight regression line to the data. (See Chapter VIII.)

In the simplest case the data appear to lie on a straight line, and what is done in such a case is to calculate from the data the equation of the best fitting linear regression line and then check the goodness of fit by the method of analysis of variance. Now the algebraic equation of a straight line is $y = ax + b$. This is equivalent to the linear regression of y on x. The constant a indicates the slope of the regression line and b the height of the line above the base. Hence a is equivalent to the regression coefficient of Section 8.i., i.e.

$$a = r\frac{\sigma_y}{\sigma_x} \quad \text{and} \quad b = \bar{y} - a\bar{x}.$$

If our data consist merely of a series of N means we have insufficient material for calculating r and σ_y. However, we know from formula (16 A) that

$$r = \frac{\frac{S(XY)}{N} - \overline{X}\overline{Y}}{\sigma_x \sigma_y}.$$

Hence $$a = r\frac{\sigma_y}{\sigma_x} = \frac{\frac{S(XY)}{N} - \overline{X}\overline{Y}}{\sigma_x^2} = \frac{S(XY) - N\overline{X}\overline{Y}}{N\sigma_x^2}. \quad (34)$$

This expression may be calculated directly from the data, since we know the exact values of x, the independent variate. The method of calculation is simple and is illustrated by the following example.

Example 35. The same test was given 10 times to a group of subjects and the mean scores obtained were as follows:

Test	1	2	3	4	5	6	7	8	9	10
Mean score	11	14	17·5	20·5	23	25·5	28	32	35·5	37·5

A graph of these means suggests that they lie roughly on a straight line. The line will be the regression of test score on order of testing so that we will call test order x, the independent variate, and mean score y, the dependent variate. We first calculate the coefficients a and b in the equation $y = ax + b$ and then proceed to examine how well a line with this equation fits the data.

Five columns of figures are constructed as in product-moment correlation. These are headed x, y, x^2, y^2 and xy respectively and each column is summed. In order to reduce the arithmetic, the y entries may be taken from an arbitrary origin, say 25 in this case. The table at the top of p. 110 shows the columns and totals obtained.

Hence $\overline{x} = 5{\cdot}5$,

$$N\overline{x}\overline{y} = \overline{x}.S(y) = 5{\cdot}5 \times -5{\cdot}5 = -30{\cdot}25,$$

$$N\overline{x}^2 = \overline{x}.S(x) = 5{\cdot}5 \times 55 = 302{\cdot}5,$$

$$a = \frac{S(xy) - N\overline{x}\overline{y}}{N\sigma_x^2} = \frac{213 + 30{\cdot}25}{385 - 302{\cdot}5} = 2{\cdot}948,$$

$$\overline{y} = -0{\cdot}55,$$

$$b = \overline{y} - a\overline{x} = -0{\cdot}55 - 2{\cdot}948 \times 5{\cdot}5 = -16{\cdot}764.$$

x	y	x^2	y^2	xy
1	−14	1	196	−14
2	−11	4	121	−22
3	−7·5	9	56·25	−22·5
4	−4·5	16	20·25	−18
5	−2	25	4	−10
6	0·5	36	0·25	3
7	3	49	9	21
8	7	64	49	56
9	10·5	81	110·25	94·5
10	12·5	100	156·25	125
55	−5·5	385	722·25	213

$N = 10$
$S(x) = 55$
$S(y) = -5{\cdot}5$
$S(x^2) = 385$
$S(y^2) = 722{\cdot}25$
$S(xy) = 213$

This value of b requires correction since y was taken from an arbitrary origin. This is done by the addition of 25, so that the true value of b is $25 - 16{\cdot}764 = 8{\cdot}236$. Hence the straight line which best fits the data is given by the equation

$$y = 2{\cdot}948x + 8{\cdot}236.$$

By substituting values of x from 1 to 10 in this equation we may calculate the best fitting theoretical values of y and compare these with the observed values, as under:

x	Observed y	Calculated y
1	11	11·184
2	14	14·132
3	17·5	17·080
4	20·5	20·028
5	23	22·976
6	25·5	25·924
7	28	28·872
8	32	31·820
9	35·5	34·768
10	37·5	37·716

It is obvious by examination that the observed values lie very closely about the calculated regression line. The significance of the fit may be tested mathematically by an analysis of the variance. (See Chapter XI.) For this purpose, the variance of y is split into two parts, that due to linear regression and that due to departures from linear regression. Since we have only mean values of y, the variance of y can be estimated with only $N-1$ degrees of freedom, in this case 9. The total sum of squares about the mean is given by $S(y^2) - \frac{[S(y)]^2}{N}$.

The part of the variance due to linear regression is given by multiplying the variance of x by a^2, i.e. it is given by

$$a^2\left[S(x^2)-\frac{[S(x)]^2}{N}\right].$$

Subtraction of this from the total sum of squares gives the part of the variance due to departures from linear regression.

We get in this example, therefore, the following analysis of variance:

	Sum of squares	D.F.	Mean squares	V.R.
1. Total	$722{\cdot}25-3{\cdot}025=719{\cdot}225$	9	—	—
2. Due to linear regression	$82{\cdot}5\times2{\cdot}948^2=716{\cdot}983$	1	716·983	2560·7
3. Due to departures from linear regression	2·242	8	0·280	

The sums of squares for lines 2 and 3 are divided by their respective number of degrees of freedom (D.F.) to obtain the mean squares or estimates of the variance. Finally the variance ratio (V.R.) is obtained by dividing the estimated variance due to linear regression by that due to departures from linear regression, giving the ratio of 2560·7. From Fisher and Yates's tables (Ref. 5) we find that for n_1 of 1 and n_2 of 8 the critical value of the variance ratio at the 1% point is 11·26. The observed value of 2560·7 is greatly in excess of this, which means that almost the whole of the variance of y is due to linear regression and the part due to departures from linear regression is negligible.

9.ix. Logarithmic curves. It is sometimes obvious that a straight regression line will not be a good fit to observed data.

Example 36. A subject was given a sensori-motor test 12 consecutive times and his scores were:

2·0, 3·3, 4·0, 4·5, 4·7, 5·0, 5·5, 5·6, 5·9, 6·0, 6·1, 6·3.

If these scores are plotted against order of testing, a line is obtained which is evidently not a straight line. The line rises fairly steeply at first and gradually flattens out, which suggests that some form of logarithmic curve might fit the data best.

The simplest forms of logarithmic curves are given by the following equations:

$$y = a\log x + b,$$
$$\log y = ax + b,$$
$$\log y = a\log x + b.$$

To discover which curve gives the best fit to the data it is necessary to plot each one of them, giving x the values of 1 to 12, and selecting the one which gives an approximately straight line. With the present data it will be found that plotting y against $\log x$ gives an almost straight line, hence we assume that the observed scores will lie very closely about a curve whose equation is $y = a\log x + b$. We have therefore to calculate the constants a and b and then examine the goodness of fit by performing an analysis of variance.

As in the previous example, we first construct and sum five columns. The first column will be values of $\log x$, which for convenience we may head X. The whole calculation is given below:

x	X ($=\log x$)	y	X^2	y^2	Xy
1	0·00	2·0	0·0000	4·00	0·0000
2	0·30	3·3	0·0900	10·89	0·990
3	0·48	4·0	0·2304	16·00	1·920
4	0·60	4·5	0·3600	20·25	2·700
5	0·70	4·7	0·4900	22·09	3·290
6	0·78	5·0	0·6084	25·00	3·900
7	0·85	5·5	0·7225	30·25	4·675
8	0·90	5·6	0·8100	31·36	5·040
9	0·95	5·9	0·9025	34·81	5·605
10	1·00	6·0	1·0000	36·00	6·000
11	1·04	6·1	1·0816	37·21	6·344
12	1·08	6·3	1·1664	39·69	6·804
	8·68	58·9	7·4618	307·55	47·268

$N = 12$
$S(X) = 8·68$
$S(y) = 58·9$
$S(X^2) = 7·4618$
$S(y^2) = 307·55$
$S(Xy) = 47·268$

$$\bar{X} = 8·68/12 = 0·7233,$$
$$N\bar{X}\bar{y} = 0·7233 \times 58·9 = 42·6024,$$
$$N\bar{X}^2 = 0·7233 \times 8·68 = 6·2782,$$
$$a = \frac{47·268 - 42·6024}{7·4618 - 6·2782} = 3·942,$$
$$\bar{y} = 58·9/12 = 4·9083,$$
$$b = 4·9083 - 3·942 \times 0·7233 = 2·0571.$$

It appears, therefore, that the original data are best fitted by a curve which has the equation $y = 3{\cdot}942 \log x + 2{\cdot}0571$. Substituting values of x from 1 to 12 in this equation we obtain the following comparison of observed and calculated values for y:

x	Observed y	Calculated y
1	2·0	2·057
2	3·3	3·244
3	4·0	3·938
4	4·5	4·431
5	4·7	4·813
6	5·0	5·125
7	5·5	5·388
8	5·6	5·617
9	5·9	5·819
10	6·0	6·000
11	6·1	6·162
12	6·3	6·311

The fit appears to be very good.

Examining this mathematically, as in the previous example, we obtain the following analysis of variance:

	Sum of squares	D.F.	Mean squares	V.R.
1. Total	$307{\cdot}55 - 289{\cdot}10 = 18{\cdot}45$	11	—	—
2. Due to linear regression	$1{\cdot}184 \times 3{\cdot}942^2 = 18{\cdot}40$	1	18·40	3680
3. Due to departures from linear regression	0·05	10	0·005	

This is a highly significant fit since the critical value of the variance ratio for $n_1 = 1$ and $n_2 = 10$ is 10·04 at the 1% point.

9.x. Polynomials. When a logarithmic curve which yields a straight regression line cannot be found, it may be necessary to fit a *polynomial* curve to observed data. Examples of such curves are:

quadratic $y = ax^2 + bx + c$,

cubic $y = ax^3 + bx^2 + cx + d$,

quartic $y = ax^4 + bx^3 + cx^2 + dx + e$, etc.

There are various ways of fitting such curves (for example, the method of least squares), but they all involve lengthy algebraical computations and will not be described in this book. (For a good account of fitting polynomials see Goulden, *Methods of Statistical Analysis*, Chapter XIV. J. Wiley and Sons, New York. 1947.)

EXERCISES ON CHAPTER IX

30. (*a*) Construct contingency tables showing the relationship between assessment *I* in Appendix E and each of the tests from *A* to *G* inclusive. The tables should be 4×3 tables, using the four categories of *I* given and dividing the subjects in each test into three groups as nearly equal in size as possible according to their scores. Suitable points of division for the groups in each test are as follows:

	(1)	(2)	(3)
A	39 and under:	40– 49:	50 and over.
B	164 and under:	165–194:	195 and over.
C	19 and under:	20– 29:	30 and over.
D	24 and under:	25– 35:	36 and over.
E	115 and under:	116–136:	137 and over.
F	24 and under:	25– 30:	31 and over.
G	33 and under:	34– 40:	41 and over.

(*b*) Calculate χ^2 and the coefficient of contingency for each table and determine in which tables there is a significant relationship.

31. (*a*) Construct 2×2 tables showing the relationship between assessment *I* and each of the tests from *A* to *G* inclusive, and also between assessment *I* and the order of merit *H*. In each case divide the assessment *I* into (1) V.G. and G.: (2) F. and P., and the test scores above and below their means as given in the answer to Exercise 3 (*b*).

(*b*) For each table calculate χ^2, using formula (29), and determine in which there is a significant relationship.

32. Investigate the association between test *E* and assessment *I* by constructing a 2×4 table and calculating χ^2 from formula (32). Divide test *E* into two categories according as the subjects are above or below the mean for the test, as given in the answer to Exercise 3 (*b*), and take assessment *I* in the four categories as given in Appendix E.

33. In a certain experiment readings had to be made from a scale which was graduated in tenths of an inch. A particular observer took 1000 readings and an analysis was made of the frequency of occurrence

of the final digits of the observations, those showing the tenth of an inch for which each reading was taken. The frequency distribution of these digits was as follows:

Digit	Frequency
0	151
1	79
2	95
3	109
4	50
5	185
6	67
7	98
8	110
9	56
	1000

There was no reason in the experiment for any particular final digits to occur more frequently than any others. Could this observer be regarded as reliable in reading the scale?

(The hypothesis to be examined here is that the observed frequency of the digits does not depart significantly from that which would be expected if the observer were completely reliable. On this hypothesis each digit is equally likely to occur, so that the expected frequency is the same for each, namely, 100. Calculate χ^2 and test its significance for 9 degrees of freedom.)

34. What is the equation of the straight line which best fits the following serial readings?

No.	1	2	3	4	5	6	7	8
Reading	5	8	$10\frac{1}{2}$	$13\frac{1}{2}$	17	21	24	26

Examine the goodness of fit by the method of analysis of variance.

Chapter X

RANKING AND THE AGREEMENT OF JUDGES

10.i. Paired comparisons: the coefficient of consistence. Frequently in experimental work the investigator wishes to rank a number of objects in order according to some quality which cannot be directly measured. He might, for instance, wish to obtain the order of preference of a number of pictures as judged by certain observers. Now it is often a matter of great difficulty for a judge to rank a whole group of objects in order of preference, but usually he can state his preference for one of a pair with ease. The method of *paired comparisons* makes use of this latter consideration by presenting the judge serially with all possible pairs of the objects to be judged, so that each object is compared with each other object separately. If there are n objects, then the number of possible pairs is $\binom{n}{2}$ which is equal to $\frac{1}{2}n(n-1)$. It may be seen that this number of pairs increases rapidly as n increases so that there is a practical limit to the number of objects that may be judged; the experiment would be too long and fatiguing with a large value of n. This may be seen from the following table showing the number of pairs for different values of n:

n	$\binom{n}{2}$	n	$\binom{n}{2}$	n	$\binom{n}{2}$	n	$\binom{n}{2}$
2	1	6	15	10	45	14	91
3	3	7	21	11	55	15	105
4	6	8	28	12	66	16	120
5	10	9	36	13	78	17	136

From the results of the comparisons of the pairs of objects a ranked order of preference of all the n objects may be worked out, but before doing this an estimate of the consistency of judgment exercised may be made. Suppose, for example, that there are 8 objects which may be designated A, B, C, D, E, F, G and H. Now if the judge prefers A to B and

B to C it follows that if he is consistent in his judgment he will prefer A to C when that pair is presented to him. If, however, he is inconsistent he might, for example, prefer A to B, B to C and C to A, which may be symbolised as $A > B > C > A$. This is known as a *circular triad*, and the number of such triads a judge gives in the whole series of paired comparisons provides an obvious measure of his degree of consistency. (It is possible for a judge to give circular polyads, i.e. four or more objects concerned, but such polyads may always be broken up into triads.)

If n is the number of objects to be judged and d is the number of circular triads shown in the whole series of comparisons, then the *coefficient of consistence*, ζ (zeta), may be calculated from one of the following expressions:

$$\text{if } n \text{ is odd,} \quad \zeta = 1 - \frac{24d}{n^3 - n}, \tag{35}$$

$$\text{if } n \text{ is even,} \quad \zeta = 1 - \frac{24d}{n^3 - 4n}. \tag{35A}$$

The next point is the method of determining d. This may be done by inspection of the series, but the method is laborious and liable to error with a long series of comparisons. The best method of finding d is as follows:

Example 37. First construct a paired comparisons table as in Table VIII. This is a square table with rows and columns headed A, B, C, ..., n and has a diagonal line drawn from the cell AA to the cell nn. In the table below $n = 8$.

With 8 objects, 28 comparisons have to be made, and the result of each comparison is entered on the table in the form of an X. An X in any cell indicates that the object at the head of the column is preferred to the object at the head of the row containing that cell. The number of X's in each column is counted and the totals entered at the foot of the table: the sum of these equals $\frac{1}{2}n(n-1)$, i.e. with 8 objects the sum of the totals is 28.

Now if there were no inconsistencies at all, the numbers at the foot of the table would all be different and would run from

TABLE VIII. *Paired comparisons table*

	A	B	C	D	E	F	G	H	
A									
B	X								
C	X	X		X		X			
D	X	X							
E	X	X	X	X					
F	X	X		X	X		X		
G	X	X	X	X	X				
H	X	X	X	X	X	X	X		
Total	7	6	3	5	3	2	2	0	28

0 to 7 and would give the ranked order of preference. This is not the case in Table VIII for there are two 3's and two 2's and 4 and 1 are missing. It follows, therefore, that there were some inconsistencies of judgment. The exact number of triads may be found by calculating

$$d = \tfrac{1}{12}n(n-1)(2n-1) - \tfrac{1}{2}\Sigma(T_x^2),$$

where T_x is any column total.

In the example, with $n = 8$,

$$d = \frac{8 \times 7 \times 15}{12} - \tfrac{1}{2}(7^2 + 6^2 + 3^2 + 5^2 + 3^2 + 2^2 + 2^2)$$

$$= 70 - 68 = 2.$$

Inspection of Table VIII shows that these two triads are $F > C > E > F$ and $F > C > G > F$. For this particular judge, $d = 2$ and $n = 8$, so that his coefficient of consistence as given by formula (35 A) is

$$\zeta = 1 - \frac{24 \times 2}{512 - 32} = 0{\cdot}90.$$

We may regard him, therefore, as having a high degree of consistency and his preference ranking of the 8 objects may be written as $ABD\overline{CE}\underline{FG}H$, C and E being assessed as equal and also F and G.

10.ii. Significance of ζ. The following table, calculated by J. W. Whitfield, gives the *maximum* values of d for significance for values of n from 6 to 20:

n	d	n	d	n	d
6	0	11	30	16	121
7	3	12	42	17	149
8	8	13	57	18	181
9	13	14	75	19	217
10	20	15	96	20	258

For ζ to be significant, d should be not greater than the value given in this table against the appropriate value of n. For values of n below 6, significance cannot be proved.

10.iii. Tied preferences. It often happens in practice that one or more pairs in a paired comparisons experiment are judged equal. There are various reasons why this may be so. A pair of objects may actually be equal in the quality being judged, or the judge may think them equal, or the judge may be incapable of making the discriminatory judgment. In the last instance he would probably show many inconsistencies in any case. Such ties are a trouble theoretically but from the experimental point of view they must be allowed.

Example 38. Table IX shows the method of dealing with such ties when they occur. This is an example where $n = 7$. Here A and C had equal preference, so the figure $\frac{1}{2}$ is written in *both* the AC and the CA cells. The same is done for the other two equal pairs, AF and DG. From the columns' totals the ranked order is $CBA\overline{DG}EF$. In this case d is 5, as calculated by the formula in the preceding section. Hence by formula (35)

$$\zeta = 1 - \frac{24 \times 5}{343 - 7} = 0{\cdot}643.$$

Since d is greater than 3, this value is not significant, so that the particular judge concerned would be regarded as inconsistent.

10.iv. The case of *m* judges: coefficient of agreement. The application of the method of paired comparisons by a single judge is chiefly of use when we know on other grounds

TABLE IX

	A	B	C	D	E	F	G	
A		X	½			½	X	
B			X					
C	½							
D	X	X	X				½	
E	X	X	X	X				
F	½	X	X	X	X		X	
G		X	X	½	X			
Total	3	5	5½	2½	2	½	2½	21

the true ranked order of the objects judged. For example, a person might profess to be able to assess intelligence from photographs. We could examine his ability to do so by photographing a number of people whose intelligence (as measured by intelligence tests) we knew and then getting the judge to rank the photographs for intelligence by the paired comparisons method. The order he chose could then be correlated with the previously known order from tests by the ranking method. (See Chapter VII.)

In practice it often occurs that we cannot directly measure the quality which we want to rank and we have to rely on personal judgments for this ranking. We might, for instance, wish to rank a number of people for the quality of initiative. In such a case as this we should be unwilling to rely on the preference of a single judge. We should get as many competent judges as possible and explain to them carefully what we mean by initiative before getting them each to rank the subjects separately by the paired comparisons method.

Suppose we had n people to be assessed and m judges. We should first examine the consistency of each judge by the method described previously. This would involve making m tables of paired comparisons and calculating the coefficient of consistency for each. In practice we should then probably omit any judge or judges who were very inconsistent since their judgments would not be reliable. Let us assume here that all m judges showed a high degree of consistency. The next point to be decided is whether or not the judges agree with one another. This may be examined by calculating the *coefficient of agreement*, u.

To do this we first construct a summed paired comparisons table. This is done by drawing up an $n \times n$ table and entering in each cell the sum of the entries in the m separate tables. For example, count the number of crosses and $\frac{1}{2}$'s in all the m AB cells in the separate tables and write the total in the AB cell of the summed table, and so on. The columns are summed as usual and the total of these sums is $\frac{1}{2}mn(n-1)$.

TABLE X. *Summed paired comparisons table: $n = 6$, $m = 5$*

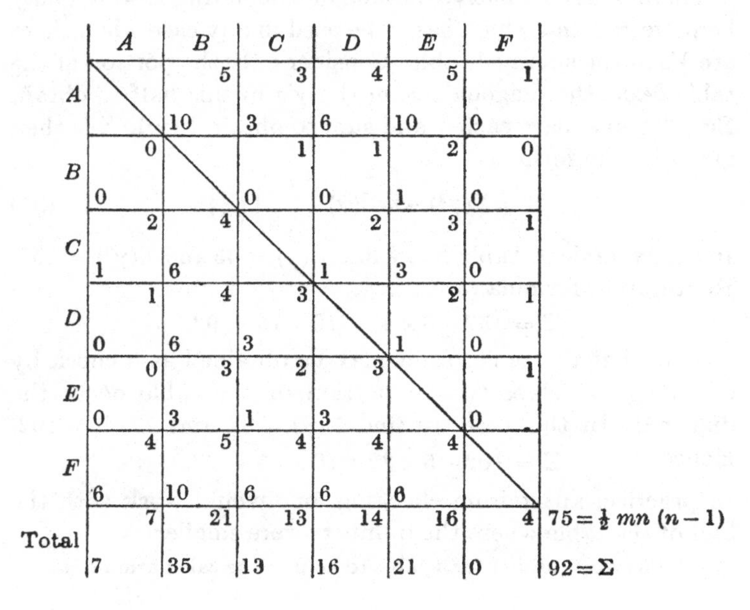

(In each cell the upper-right figure is given first and the lower-left figure second.)

	A	B	C	D	E	F	
A		5 / 10	3 / 3	4 / 6	5 / 10	1 / 0	
B	0 / 0		1 / 0	1 / 0	2 / 1	0 / 0	
C	2 / 1	4 / 6		2 / 1	3 / 3	1 / 0	
D	1 / 0	4 / 6	3 / 3		2 / 1	1 / 0	
E	0 / 0	3 / 3	2 / 1	3 / 3		1 / 0	
F	4 / 6	5 / 10	4 / 6	4 / 6	4 / 6		
Total	7 / 7	21 / 35	13 / 13	14 / 16	16 / 21	4 / 0	$75 = \frac{1}{2}mn(n-1)$ / $92 = \Sigma$

Example 39. Table X shows a summed paired comparisons table for 6 objects judged by 5 observers, so that $n = 6$ and $m = 5$. We shall deal first with the case where there are no $\frac{1}{2}$'s in the summed table in order to illustrate two methods of calculation. The number written in the top right-hand corner of each cell is the total of the entries in the 5 corresponding cells in the separate tables (which are not reproduced). Let this entry for any cell be symbolised by γ (gamma). Then for each cell we calculate $\frac{1}{2}\gamma(\gamma-1)$ and write the answer in the bottom left-hand corner of each cell. If we add all these values together we obtain $\Sigma\binom{\gamma}{2}$, which we may call Σ. Then u, the coefficient of agreement, is given by the formula

$$u = \frac{2\Sigma}{\binom{m}{2}\binom{n}{2}} - 1 = \frac{8\Sigma}{mn(m-1)(n-1)} - 1. \qquad (36)^*$$

From the data in Table X we obtain $\Sigma = 92$. Hence in this example $u = \frac{8 \times 92}{5 \times 6 \times 4 \times 5} - 1 = 0{\cdot}227$.

There is an alternative method of calculating Σ which may be preferred and which has to be used in any case when there are $\frac{1}{2}$'s in the summed table. Consider only the portion of the table *below* the diagonal. Sum the γ's in this half to obtain $\Sigma(\gamma)$. Then square each γ and sum to obtain $\Sigma(\gamma^2)$. Σ is then given by the formula

$$\Sigma = \Sigma(\gamma^2) - m\Sigma(\gamma) + \binom{m}{2}\binom{n}{2}. \qquad (37)$$

In the example in Table X, we find $\Sigma(\gamma) = 43$ and $\Sigma(\gamma^2) = 157$. Hence, from formula (37),

$$\Sigma = 157 - 5 \times 43 + 10 \times 15 = 92.$$

Note that the same result may be obtained as a check by confining ourselves to the portion of the table *above* the diagonal. In this case we find $\Sigma(\gamma) = 32$ and $\Sigma(\gamma^2) = 102$. Hence

$$\Sigma = 102 - 5 \times 32 + 10 \times 15 = 92.$$

In practice, apart from checking, one would work with the half of the table where the numbers were smaller.

* See Appendix H for a table to help in the calculation of u.

10.v. Significance of u. The significance of u may be tested by calculating χ^2. Making a correction for continuity (similar to Yates's correction), we have in this case the following expression for χ^2:

$$\chi^2 = \left\{\Sigma - 1 - \tfrac{1}{2}\binom{m}{2}\binom{n}{2}\frac{m-3}{m-2}\right\}\frac{4}{m-2}. \qquad (38)^*$$

The number of degrees of freedom, ν (nu), corresponding to this value of χ^2 is given by

$$\nu = \frac{\binom{n}{2} m(m-1)}{(m-2)^2}. \qquad (39)^*$$

The significance of χ^2 is then examined by calculating $\sqrt{(2\chi^2)} - \sqrt{(2\nu - 1)}$. If this expression is greater than 1·65, χ^2 and consequently u is significant.

In the example in 10.iv we have $n = 6$, $m = 5$ and $\Sigma = 92$. From formula (38)

$$\chi^2 = \{92 - 1 - \tfrac{1}{2}(15 \times 10)\tfrac{2}{3}\}\tfrac{4}{3} = 54\tfrac{2}{3}.$$

From formula (39)

$$\nu = \frac{15 \times 5 \times 4}{3 \times 3} = 33\tfrac{1}{3}.$$

Hence $$\sqrt{(2\chi^2)} - \sqrt{(2\nu - 1)} = \sqrt{(109{\cdot}3)} - \sqrt{(65{\cdot}6)}$$
$$= 10{\cdot}46 - 8{\cdot}10 = 2{\cdot}36.$$

This is greater than 1·65, therefore u is significant.

From the table of the normal curve (see Ref. 1, Appendix, Table 2), it may be seen that the probability of getting a value of χ^2 as large or larger than that obtained is $P = 0{\cdot}009$.

(*Note.* For exact probabilities for values of m from 3 to 6 inclusive and small values of n, see Kendall, vol. I, Tables 16·11–16·14, Ref. 6.)

10.vi. Combination of several rankings. Suppose we have obtained rankings of n objects by m judges, either directly or by the method of paired comparisons, and we wish

* See Appendix I for values of ν and Appendix J for a table to help in the calculation of χ^2.

to combine these separate rankings into a single ranking which may be taken as the best representation of the consensus of judgment. The simplest and best way of doing this is to sum the m rankings for each object and re-rank the n totals thus obtained.

Example 40. Suppose 3 judges ranked 8 objects as follows:

	A	B	C	D	E	F	G	H
Judge 1	4	2	1	7	6	3	5	8
Judge 2	7	2	1	6	4	5	3	8
Judge 3	7	4	2	6	5	3	1	8
Sum	18	8	4	19	15	11	9	24
Combined rank	6	2	1	7	5	4	3	8

The last row is the combined rank order obtained by ranking the sums in the row above.

Sometimes the sums of rankings for two objects may be equal, for example $1+1+3 = 5$ and $2+2+1 = 5$. These two objects would be tied in the combined ranking, but if for a particular purpose we wished to avoid ties, we should take as the smaller of the pair the one which gave the smaller sum of squares of its constituents. Thus the sum of the squares of 1, 1 and 3 is equal to 11, whilst the sum of the squares of 2, 2 and 1 is equal to 9, so that we should take the latter as of lesser rank than the former. If, however, there were tied rankings in the separate judgments, we should have to allow ties in the combined ranking.

10.vii. Coefficient of concordance. The degree of resemblance between m rankings of the same n objects may be estimated by calculating the *coefficient of concordance*, W. To do this, first sum the m ranks for each object as in 10.vi. The total of these sums is $\frac{1}{2}mn(n+1)$ and their mean is $\frac{1}{2}m(n+1)$. Take the deviation of each sum from this mean and square it. Add the squares to obtain a total sum of squares which we will call S. S is then the variance of the sums of rankings. Now if the concordance amongst the judges were perfect, these sums would be m, $2m$, $3m$, ..., nm, and their variance would be $\frac{1}{12}m^2(n^3-n)$. The coefficient W is then given by the ratio of the observed variance S to the variance for perfect concordance, i.e.

$$W = \frac{12S}{m^2(n^3-n)}. \qquad (40)^*$$

In Example 40, $m = 3$ and $n = 8$. The sums of ranks were 18, 8, 4, 19, 15, 11, 9 and 24. These total to 108 with a mean of $13\frac{1}{2}$. Then S is equal to

$$(18-13\tfrac{1}{2})^2+(8-13\tfrac{1}{2})^2+\ldots+(24-13\tfrac{1}{2})^2.$$

The total of this is 310. Hence, from formula (40),

$$W = \frac{12\times 310}{9(512-8)} = 0{\cdot}82.$$

W varies between 0 and 1 and is unity only when all the rankings are identical.

W is related to the average Spearman ρ between all possible pairs of rankings, the relationship being given by

$$\rho_{\text{av.}} = \frac{mW-1}{m-1}.$$

Formula (40) holds only when there are no tied ranks. When ties are present it needs modification. Consider any one row of rankings. If there are t ranks tied in any one set, calculate $\frac{1}{12}(t^3-t)$. Do this for each set of ties and add for the whole row, obtaining $\frac{1}{12}\Sigma(t^3-t)$. Call this total T_1 for the first row, T_2 for the second and so on. Then if there are m rows, add the m values of T to obtain $\Sigma(T)$. Then the modified form of formula (40) becomes

$$W = \frac{S}{\frac{1}{12}m^2(n^3-n)-m\Sigma(T)}. \qquad (40\text{ A})$$

The following small table is of help in calculating the T's:

t	$\frac{1}{12}(t^3-t)$
2	$\frac{1}{2}$
3	2
4	5
5	10
6	$17\frac{1}{2}$

* See Appendix K for a table of $m^2(n^3-n)/12$.

Example 41. Calculate the value of W for the following data, showing 3 rankings of the same 10 objects:

Object	A	B	C	D	E	F	G	H	J	K
Ranking 1	1	2	3	$4\frac{1}{2}$	$4\frac{1}{2}$	6	$7\frac{1}{2}$	$7\frac{1}{2}$	9	10
Ranking 2	1	$2\frac{1}{2}$	$2\frac{1}{2}$	$4\frac{1}{2}$	$4\frac{1}{2}$	$6\frac{1}{2}$	$6\frac{1}{2}$	8	$9\frac{1}{2}$	$9\frac{1}{2}$
Ranking 3	1	2	$4\frac{1}{2}$	$4\frac{1}{2}$	$4\frac{1}{2}$	$4\frac{1}{2}$	8	8	8	10
Sums	3	$6\frac{1}{2}$	10	$13\frac{1}{2}$	$13\frac{1}{2}$	17	22	$23\frac{1}{2}$	$26\frac{1}{2}$	$29\frac{1}{2}$

The sums total 165 with a mean of $16\frac{1}{2}$. Subtracting $16\frac{1}{2}$ from each sum, squaring and adding we get $S = 691$.

Now in the first row there are 2 ties with 2 members each, so that $T_1 = \frac{1}{2} + \frac{1}{2} = 1$. Similarly, $T_2 = 2$ and $T_3 = 7$. Therefore $\Sigma(T) = 10$. Then, from formula (40 A),

$$W = \frac{12 \times 691}{9(1000-10) - 3 \times 10} = 0{\cdot}970.$$

10.viii. Significance of W. The significance of W may be tested as for Fisher's z (Section 8.iii) by writing

$$z = \tfrac{1}{2}\log_e \frac{(m-1)\,W}{1-W}. \tag{41}$$

The appropriate degrees of freedom for consulting the z table (Fisher and Yates, Ref. 5, Table V) are

$$\left.\begin{aligned} \nu_1 &= (n-1) - \frac{2}{m}, \\ \nu_2 &= (m-1)\,\nu_1. \end{aligned}\right\} \tag{42}$$

When ties are present, the same test may be used provided $\Sigma(T)$ is small compared with $\frac{1}{12}m^2(n^3-n)$. If a large part of $\Sigma(T)$ is provided by the tied ranks from one judge, we might regard him as being relatively incapable of making the judgments and so omit his rankings. Great care has to be exercised, however, in omitting any experimental data—full justification has to be shown. A number of judges may have a high degree of concordance in their judgments and yet be wrong. An outstanding case might disagree with the others and yet be right.

An alternative form of formula (41) using logs to the base 10 instead of natural logarithms is

$$z = 1{\cdot}1513[\log(m-1)\,W - \log(1-W)]. \tag{41 A}$$

Example 42. Examine the significance of the value of W obtained in Example 40. Here $m = 3$, $n = 8$, $W = 0{\cdot}82$.

Hence
$$\begin{aligned} z &= 1{\cdot}1513[\log(2\times 0{\cdot}82) - \log(1-0{\cdot}82)] \\ &= 1{\cdot}1513(0{\cdot}2148 - \bar{1}{\cdot}2553) \\ &= 1{\cdot}1513 \times 0{\cdot}9595 \\ &= 1{\cdot}1047. \end{aligned}$$

$\nu_1 = 6\frac{1}{3}$ and $\nu_2 = 12\frac{2}{3}$. From Table V in Fisher and Yates (Ref. 5) we find for $\nu_1 = 6$ and $\nu_2 = 13$ a value of z of $0{\cdot}5350$ for the 5 % point of significance. The calculated value is much greater than this and is therefore significant.

Using formula (41) we get

$$\frac{(m-1)W}{1-W} = \frac{1{\cdot}64}{0{\cdot}18} = 9{\cdot}11,$$

$$\log_e 9{\cdot}11 = 2{\cdot}2094,$$

$$z = \tfrac{1}{2}\log_e 9{\cdot}11 = 1{\cdot}1047.$$

(*Note.* The methods described in this chapter are mainly due to M. G. Kendall; for their derivation, etc., see his *Rank Correlation Methods*, Griffin and Co., 1948.)

EXERCISES ON CHAPTER X

35. In a paired comparisons experiment, a subject was required to rank 7 photographs, labelled A to G, in order of the intelligence of expression portrayed. All possible pairs were presented in random order and the 21 judgments made were as follows (recorded here in regular order for convenience):

$A>B$	$B>D$	$C<G$
$A>C$	$B>E$	$D>E$
$A>D$	$B<F$	$D>F$
$A<E$	$B<G$	$D<G$
$A>F$	$C>D$	$E>F$
$A<G$	$C>E$	$E<G$
$B>C$	$C>F$	$F<G$

Construct a paired comparisons table from these data.

(*a*) Calculate d, the number of circular triads, and find out from the table which are the triads.

(*b*) Calculate ζ, the coefficient of consistence.

(*c*) What is the ranked order of the 7 photographs?

(*d*) What conclusions would you draw from these findings?

36. The experiment in the previous exercise was made on 6 subjects and the results are shown in the summed paired comparisons table below:

	A	*B*	*C*	*D*	*E*	*F*	*G*	
A		0	0	0	3	2	6	
B	6		2	2	1	4	6	
C	6	4		1	1	3	6	
D	6	4	5		1	0	6	
E	3	5	5	5		0	6	
F	4	2	3	6	6		6	
G	0	0	0	0	0	0		
	25	15	15	14	12	9	36	126

(*a*) Calculate u, the coefficient of agreement, and examine its significance.

(*b*) What conclusions do you now draw about the photographs?

37. Ten men in a factory were ranked in order of their degree of possession of initiative by each of 4 foremen. These were the rankings:

Workman	1	2	3	4	5	6	7	8	9	10
Ranking 1	4	3	9	10	1	8	2	6	5	7
Ranking 2	4	1	10	9	2	5	3	8	6	7
Ranking 3	7	4	$8\frac{1}{2}$	$8\frac{1}{2}$	$1\frac{1}{2}$	5	$1\frac{1}{2}$	6	3	10
Ranking 4	5	2	8	9	2	7	2	10	4	6

Calculate W, the coefficient of concordance between the rankings and examine its significance.

Chapter XI

INTRODUCTION TO ANALYSIS OF VARIANCE

11.1. Fundamental nature. The technique of analysis of variance (which is really a misnomer, as may be seen later) was developed by R. A. Fisher. It has since been so widely used and elaborated that the literature on the subject is enormous and discouraging to the elementary student. The object of this chapter is to explain the fundamental nature of the method and to give some simple illustrations of its use.

Essentially, analysis of variance provides a test of the *homogeneity* of a set of data. By homogeneity we mean that all the observations could have been drawn from the same population, which we may call the parent population, this population being normally distributed and having a certain mean and variance. In any experiment there may be several factors at work, each of which may cause a certain amount of variability in the observations made. Thus if several varieties of potatoes are sown in different types of soil and treated with different sorts of manure, the variation in the various yields of tubers may be affected by variety of seed, by variety of soil or by variety of manure. Analysis of variance enables us to discover how much of the total variability is due to each factor and a comparison of these contributory amounts of variation provides a test of the homogeneity of the observations. If the data are shown not to be homogeneous, then the factor or factors causing the heterogeneity may be isolated and the relative amounts of their effects discovered. It should be emphasised at the outset that for this analysis to be possible, it is essential that the experiment be very carefully planned so as to justify the assumptions on which the method of variate analysis is based. These assumptions will be pointed out later.

The method derives from the fact that variance is an

additive quantity. Suppose X and Y are two independent, normally distributed variables. (By independent is meant uncorrelated.) Then the variance of X is $\frac{1}{N}S(X-\bar{X})^2$ and that of Y is $\frac{1}{N}S(Y-\bar{Y})^2$, where N is the number in a sample of each variable. In a similar way, the variance of $(X+Y)$ will be

$$\frac{1}{N}S\{(X+Y)-(\overline{X+Y})\}^2 = \frac{1}{N}S(X+Y-\bar{X}-\bar{Y})^2.$$

Rearranging and expanding we have

$$\frac{1}{N}S\{(X-\bar{X})+(Y-\bar{Y})\}^2$$

$$= \frac{1}{N}S\{(X-\bar{X})^2+(Y-\bar{Y})^2+2(X-\bar{X})(Y-\bar{Y})\}$$

$$= \frac{1}{N}S(X-\bar{X})^2+\frac{1}{N}(Y-\bar{Y})^2+\frac{2}{N}(X-\bar{X})(Y-\bar{Y}).$$

The last term on the right is twice the co-variance of X and Y and since the two variables are uncorrelated this must be equal to zero. Hence the variance of

$$(X+Y) = (\text{variance of } X) + (\text{variance of } Y).$$

The same sort of reasoning may be applied to three or more variables.

If therefore we have a set of observations each of which may be regarded as falling into several independent classifications, the total variance may be analysed into several variances, one for each of the separate classifications. Further, if the classes are really independent and the data are homogeneous, then the variance in each classification is an independent estimate of the parent variance, and these separate estimates will all be equal within the limits of sampling error. If they are not equal, then the data are not homogeneous and we come to the conclusion that the data in different classes are drawn from different populations having different means. The important point to remember in applying the test for homogeneity (how this is done is explained below) is that three assumptions are made:

(1) that the data are normally distributed;

(2) that the separate estimates of variance are independent;

(3) that the variance in each class is the same.

11.ii. One-way classification. In Section 5.iii we examined the probability that two sample means came from the same population, using the t test. If we had three or more samples it would be possible to apply the t test to each pair of means, but the homogeneity of the samples can be tested more easily by the analysis of variance.

Example 43. Numbers of leaves were taken from each of half a dozen trees and their lengths measured. The following are the measurements in millimetres:

Tree	Length
1	82, 87, 86, 90, 81, 84
2	85, 84, 91, 92, 88
3	92, 90, 84, 86, 88, 93, 89, 90
4	80, 86, 87, 81, 82, 82
5	87, 86, 88, 90, 85, 86, 87
6	90, 86, 84, 85, 85, 86, 87, 84, 87

What we wish to know is: can all these leaves be regarded as having come from the same species of tree?

Now if X represents the length of any leaf and we lump all the samples together to form one group of number N, then the variance of the whole group is given as usual by

$$\sigma^2 = \frac{S(X-\bar{X})^2}{N-1},$$

since we are *estimating* the variance of the parent population.

The expression $S(X-\bar{X})^2$ is the sum of the squares of deviations from the mean and will be referred to for convenience as s.o.s. This sum of squares may be analysed into two parts, each one of which provides an independent estimate of the parent variance.

Let $\bar{X}$ be the general mean of the whole group and $\bar{X}_s$ any sample mean. Then the total s.o.s. may be written

$$S(X-\bar{X})^2 = S(X-\bar{X}_s+\bar{X}_s-\bar{X})^2.$$

Grouping and expanding the right-hand side we get

$$S\{(X-\bar{X}_s)+(\bar{X}_s-\bar{X})\}^2$$
$$= S(X-\bar{X}_s)^2+S(\bar{X}_s-\bar{X})^2+2S(X-\bar{X}_s)(\bar{X}_s-\bar{X}).$$

The product term on the right equals zero, hence we may write

$$S(X-\bar{X})^2 = S(X-\bar{X}_s)^2+S(\bar{X}_s-\bar{X})^2.$$

The first part on the right is the s.o.s. of the observations in each sample about their respective means: the second is the s.o.s. of the sample means about the general mean. To calculate these sums of squares we make use of the identity

$$S(X-\bar{X})^2 = S(X^2)-[S(X)]^2/N.$$

First we find $S(X)$ and $S(X^2)$ for each sample. These may conveniently be arranged with the sample means as shown in Table XI.

TABLE XI

Sample	No. in sample, N_s	$S(X)$	$S(X^2)$	$\bar{X}_s$
1	6	510	43,406	85
2	5	440	38,770	88
3	8	712	63,430	89
4	6	498	41,374	83
5	7	609	52,999	87
6	9	774	66,592	86
Totals	41 = N	3543	306,571	86·41 = $\bar{X}$

Note that for sample 1, $S(X^2)$ is given by

$$82^2+87^2+86^2+90^2+81^2+84^2 = 43{,}406,$$

and similarly for the other samples.

The total s.o.s. is then

$$306571-3543^2/41$$
$$= 306571-306167{\cdot}05$$
$$= 403{\cdot}95.$$

The s.o.s. of the sample means about the general mean, i.e. $S(\bar{X}_s-\bar{X})^2$, is obtained by squaring $S(X)$ for each sample, dividing each square by N_s, the number in that sample, adding the six results and subtracting $[S(X)]^2/N$. Thus we have

$$S(\bar{X}_s - \bar{X})^2 = \frac{510^2}{6} + \frac{440^2}{5} + \frac{712^2}{8} + \frac{498^2}{6} + \frac{609^2}{7} + \frac{774^2}{8} - \frac{3543^2}{41}$$
$$= 306319 - 306167{\cdot}05$$
$$= 151{\cdot}95.$$

The S.O.S. of sample observations about sample means is obtained by subtracting $[S(X)]^2/N_s$ from $S(X^2)$ for each sample and adding the six results. We have, therefore,

$$S(X - \bar{X}_s)^2 = (43406 - 510^2/6) + (38770 - 440^2/5)$$
$$+ (63430 - 712^2/8) + (41374 - 498^2/6)$$
$$+ (52999 - 609^2/7) + (66592 - 774^2/9)$$
$$= 56 + 50 + 62 + 40 + 16 + 28$$
$$= 252.$$

We now have the necessary sums of squares. To obtain the estimates of variance these have to be divided by the respective number of degrees of freedom (D.F.). The total number of D.F. is $41 - 1 = 40$. If there are P samples, the D.F. $= P - 1 = 5$ in this case. The D.F. from individuals in the samples about their sample means $= S(N_s - 1) = N - P = 35$. The complete analysis of variance then takes the following tabular form:

	S.O.S.	D.F.	Mean square
Between samples	151·95	5	30·39
Within samples	252	35	7·20
Total	403·95	40	10·099

It will be noted that the first two columns of figures are additive but not the last column, which gives the mean squares or estimates of the parent variance. In other words, we have actually analysed the total S.O.S. and not the variance. The first two variances in the last column are independent estimates and so may be compared. The variance calculated from the total S.O.S. is not independent since it includes the other two.

There are now two ways of testing the homogeneity of the samples. First we may calculate

$$z = \frac{1}{2}\left\{\log_e \frac{S(\bar{X}_s - \bar{X})^2}{P-1} - \log_e \frac{S(X - \bar{X}_s)^2}{N-P}\right\},$$

and consult Table V in Fisher and Yates's Tables for $n_1 = 5$ and $n_2 = 35$. In this case $z = 0{\cdot}7200$. From the table for $n_1 = 5$ and $n_2 = 30$ we see a value of z of 0·4648 for $P = 0{\cdot}05$. Since the calculated z is much bigger than the one in the tables, it means that the data depart significantly from homogeneity.

Alternatively, and more simply, we may calculate the *variance ratio*, V.R., by dividing the variance between samples by the variance within samples and again consulting Table V in Fisher and Yates. In this case V.R. = 30·39/7·2 = 4·22. In the table for $n_1 = 5$ and $n_2 = 30$ we find a value for the V.R. of 2·53 for $P = 0{\cdot}05$, again showing that the calculated V.R. would occur by chance very much less often than 1 in 20 times. Hence the data depart significantly from homogeneity.

The above example has been worked out in full to show exactly what is being done. In practice the arithmetic may be reduced. First we take a working mean of, say, 85 and reduce the original data to the following:

Tree	Length (from 85)
1	−3, 2, 1, 5, −4, −1
2	0, −1, 6, 7, 3
3	7, 5, −1, 1, 3, 8, 4, 5
4	−5, 1, 2, −4, −3, −3
5	2, 1, 3, 5, 0, 1, 2
6	5, 1, −1, 0, 0, 1, 2, −1, 2

From these data Table XI becomes Table XI A.

TABLE XI A

Sample	N_s	$S(X)$	$S(X^2)$	$\bar{X}_s$
1	6	0	56	0
2	5	15	95	3
3	8	32	190	4
4	6	−12	64	−2
5	7	14	44	2
6	9	9	37	1
Totals	$41 = N$	58	486	$1{\cdot}41 = \bar{X}$

From this, total

$$\text{s.o.s.} = 486 - 58^2/41 = 486 - 82{\cdot}05 = 403{\cdot}95.$$

For the sum of squares between samples we make use of the identity

$$S(\bar{X}_s - \bar{X})^2 = S[S(X)\,\bar{X}_s] - [S(X)]^2/N.$$

In this case we have

$$\begin{aligned}\text{s.o.s. between samples} \\ &= (0\times 0) + (15\times 3) + (32\times 4) + (-12\times -2) \\ &\quad + (14\times 2) + (9\times 1) - 58\times 58/41 \\ &= 0 + 45 + 128 + 24 + 28 + 9 - 82{\cdot}05 \\ &= 151{\cdot}95.\end{aligned}$$

The s.o.s. of observations about their respective sample means need not be calculated directly but may be obtained by subtracting the s.o.s. between samples from the total s.o.s. This may be entered in the analysis of variance table as the 'remainder'. The table then takes this form:

	S.O.S.	D.F.	Mean square	V.R.
Between samples	151·95	5	30·39	4·22
Remainder	252	35	7·20	—
Total	403·95	40	10·099	

11.iii. *Interpretation of analysis.* Having decided from our analysis that the data are not homogeneous we have to ask what this signifies. To begin with, it means that the six examples are not all drawn from the same parent population. The sample means vary from 83 to 89 and we may decide fairly definitely that the sample with the smallest mean length was taken from a different species of tree from that with the largest mean length. How much further may we go? This may be decided by calculating the standard error of the difference between sample means. Now we have three estimates of variance. If the data were homogeneous, the best estimate would be the one based on the largest number of D.F., i.e. that obtained from the total s.o.s. However, the data are not homogeneous, so that the best estimate of the variance for calculating the standard error is the 'remainder' variance, since from this the variance due to differences between sample means has been eliminated. This variance is 7·20, so that the standard error of a sample of size N drawn from a population of estimated variance 7·20 will be $\sqrt{(7{\cdot}20/N)} = 2{\cdot}683/\sqrt{N}$.

The standard error of the difference between two means of samples of N_1 and N_2 respectively will be

$$2{\cdot}683\sqrt{\left(\frac{1}{N_1}+\frac{1}{N_2}\right)} \quad \text{(see Chapter v).}$$

The difference between the means of samples 3 and 4 in the previous example is $89-83 = 6$ and the numbers in the samples are 8 and 6. Hence

S.e. of the difference

$$= 2{\cdot}683\sqrt{(1/8+1/6)} = 2{\cdot}683 \times 0{\cdot}54$$
$$= 1{\cdot}449.$$

Difference/S.e. of difference

$$= 6/1{\cdot}449 = 4{\cdot}14.$$

Normally we should infer from this, if we had only two samples, that this difference between the means is definitely significant. However, a little care is necessary in coming to this conclusion. We have 6 samples so that we might compare 15 different pairs of means. Above we chose the biggest difference so that we ought to apply a stricter test of significance than that of $P = 0{\cdot}05$. It seems reasonable in this case to demand that the probability of a difference of the observed size arising by chance should be 1 in 20×15, i.e. 1 in 300. This corresponds roughly to a case where a difference is 2·9 times its standard error, as may be seen from the table of the probability integral. Hence we may take it in this case that any difference between means that is 2·9 times its standard error, or greater, is significant. Applying this test to all possible pairs in the example we obtain the results shown in the table on p. 137.

The order of the samples arranged according to mean value is 3, 2, 5, 6, 1, 4. From the table it may be seen that only samples 3 and 4 and 2 and 4 are significantly different. We should suspect, therefore, that samples 2 and 3 came from a

Samples	Difference	S.e. of diff.	Diff./S.e.
1 and 2	3	1·625	1·8
1 and 3	4	1·449	2·8
1 and 4	2	1·549	1·3
1 and 5	2	1·493	1·3
1 and 6	1	1·414	0·7
2 and 3	1	1·530	0·7
2 and 4	5	1·625	3·1 Significant
2 and 5	1	1·571	0·6
2 and 6	2	1·497	1·3
3 and 4	6	1·449	4·1 Significant
3 and 5	2	1·388	1·4
3 and 6	3	1·304	2·3
4 and 5	4	1·493	2·7
4 and 6	3	1·414	2·1
5 and 6	1	1·352	0·7

different species of tree from sample 4 but which of the other samples came from which species cannot be deduced from the present data. There is an indication that with larger samples the differences between the means of 3 and 6 and 3 and 1 might appear as significant, and also that between 4 and 5. We might suspect then that samples 3, 2 and 5 came from one species of tree and samples 6, 1 and 4 from another, but we could not prove this from the present data. For this we should need larger samples or some other means of identification, such as shape.

11.iv. Two-way classification. In the last example, the data all belonged to a single family, leaf-lengths. It frequently happens that a set of observations may be regarded as belonging to two families at the same time.

Example 44. In an experiment on the effects of temperature conditions on human performance, 8 practised subjects were given a sensori-motor test in each of 4 temperature conditions. Since the subjects were well practised, the order in which the tests were done was unimportant. The tests were randomised amongst the subjects, so that for each condition there were equal numbers of first testing, second testing, third testing and fourth testing. The scores in the tests are shown below.

Table XII

		Subjects (*A*)								
		1	2	3	4	5	6	7	8	Totals
Temperatures (*B*)	1	76	80	79	90	85	101	94	83	688
	2	75	81	77	90	86	98	93	85	685
	3	76	78	76	91	82	98	92	83	676
	4	68	75	72	85	82	90	82	77	631
Totals		295	314	304	356	335	387	361	328	2680

Call the number of subjects N_a and the number of conditions N_b. In such a case as this the total s.o.s. may be analysed into three parts:

(i) the s.o.s. of A means about the general mean, i.e. $S(\bar{X}_a - \bar{X})^2$, with $N_a - 1$ D.F.;

(ii) the s.o.s. of B means about the general mean, i.e. $S(\bar{X}_b - \bar{X})^2$, with $N_b - 1$ D.F.;

(iii) a remainder obtained by subtracting (i) and (ii) from the total s.o.s. This has $(N_a - 1)(N_b - 1)$ D.F.

The total s.o.s., with $N_a N_b - 1$ D.F., is calculated as in the previous example by summing the squares of all the readings and subtracting $[S(X)]^2/N$. For convenience and to reduce the arithmetic the data in Table XII may be written down from a working mean of 84 (since $2680/32 = 83{\cdot}75$). This yields Table XII A.

Table XII A

		Subjects (*A*)								
		1	2	3	4	5	6	7	8	Totals
Temperatures (*B*)	1	−8	−4	−5	6	1	17	10	−1	16
	2	−9	−3	−7	6	2	14	9	1	13
	3	−8	−6	−8	7	−2	14	8	−1	4
	4	−16	−9	−12	1	−2	6	−2	−7	−41
Totals		−41	−22	−32	20	−1	51	25	−8	−8

The total s.o.s.

$$= (-8)^2 + (-4)^2 + (-5)^2 + \ldots + (-2)^2 + (-7)^2 - (-8)^2/32$$

$$= 2042 - 2 = 2040.$$

Great care should be taken with this part of the calculation as there is no simple check on the arithmetic.

To obtain $S(\bar{X}_a - \bar{X})^2$, square and sum the 8 totals for the A grouping, divide by 4 (since there are 4 readings in each), and subtract $[S(X)]^2/N$.

Hence

$$\begin{aligned} S(\bar{X}_a - \bar{X})^2 &= \tfrac{1}{4}[(-41)^2 + (-22)^2 + (-32)^2 + 20^2 + (-1)^2 \\ &\quad + 51^2 + 25^2 + (-8)^2] - 2 \\ &= 6880/4 - 2 \\ &= 1718. \end{aligned}$$

In a similar way, $S(\bar{X}_b - \bar{X})^2$ is obtained by squaring and summing the 4 totals for the B groups, dividing by 8 and subtracting $[S(X)]^2/N$. Hence

$$\begin{aligned} S(\bar{X}_b - \bar{X})^2 &= \tfrac{1}{8}[16^2 + 13^2 + 4^2 + (-41)^2] - 2 \\ &= 2122/8 - 2 \\ &= 263{\cdot}25. \end{aligned}$$

We may now construct the analysis table as below, obtaining the 'remainder' s.o.s. by subtraction.

	S.O.S.	D.F.	Mean square	V.R.
Between subjects	1718	7	245·4	87·6
Between temperatures	263·25	3	87·75	31·3
Remainder	58·75	21	2·80	—
Total	2040	31	65·8	

In this table the first three estimates of the parent variance, given in the mean square column, are independent. The V.R. is obtained in each case by calculating the ratio of the first two variances to the remainder variance. Reference to Fisher and Yates's Table V shows that both these ratios are highly significant, so that we reject the hypothesis that the data are homogeneous. The variance between subjects is very large, showing that there are highly significant differences in ability between the subjects, but we are not interested in this. What does interest us is the effect of temperature conditions. The means for these conditions, obtained by dividing the B totals in Table XII by 8, are 86, 85·625, 84·5 and 78·875. Since the data are not homogeneous we take the remainder variance as the best estimate of the parent variance, so that the standard error of a mean of N individuals drawn at random will be $\sqrt{(2{\cdot}80/N)} = 1{\cdot}673/\sqrt{N}$.

Now we have 6 possible pairs of means to compare, hence it is reasonable to use a level of significance equal to 1 in 20×6, i.e. 1 in 120. This corresponds to a ratio between a difference and its standard error of about 2·6. Since there are 8 readings in each temperature mean, the standard error of the difference between the means of two samples is $1{\cdot}673\sqrt{(\frac{1}{8}+\frac{1}{8})} = {\cdot}84$. Hence means differing by more than 2·2 will be significantly different, i.e. the mean of the fourth temperature condition is significantly different from the means of the other three conditions, but these three do not differ significantly from one another.

11.v. Three-way classification. In the previous example, the remainder S.O.S., which would be symbolised as $S(X - X_a - X_b + \bar{X})^2$, is sometimes referred to as an *interaction*. When we have a set of data which can be classified in three ways, A, B and C, there will be three such terms, one for the interaction of A and B, one for A and C and the third for B and C. The total S.O.S. in such a case may therefore be analysed into seven parts. First there will be three parts given by $S(\bar{X}_a - \bar{X})^2$, $S(\bar{X}_b - \bar{X})^2$ and $S(\bar{X}_c - \bar{X})^2$; these show the effects of A, B and C separately and are usually known as the *main effects*. Next will be the three interaction terms mentioned above. Interactions involving two main effects are known as *first-order interactions*. Finally, the seventh part in the analysis is a remainder term. This term is in fact the interaction of all three main effects and is known as a *second-order interaction*. The number of degrees of freedom involved in interactions is always the product of the D.F. of the component main effects, so that if the numbers of groupings in the classes A, B and C are p, q and r respectively, the D.F. for interaction AB will be $(p-1)(q-1)$, for AC $(p-1)(r-1)$ and for BC $(q-1)(r-1)$, whilst for the second-order interaction ABC the D.F. will be $(p-1)(q-1)(r-1)$. The total D.F. are $(pqr-1)$ and the student may easily verify by algebra that this is equal to the sum of the D.F. of the seven parts in the analysis.

The calculation of the various sums of squares is similar

in nature to that of the previous example but a little more complicated. The method is fully illustrated in the following example.

Example 45. In a certain psychological experiment the apparatus consisted of a dial having a rotating needle and a number of graduations round the circumference. The needle could be rotated at 3 different speeds and the dial displayed under 3 different intensities of illumination. Subjects in this experiment were required to make a certain reaction each time the needle reached a graduation on the dial. Since there are nine combinations of speed and illumination, each subject had 9 tasks to perform. The order of performance was randomised for different subjects. Table XIII below gives the number of correct reactions made in each of the 9 tasks by 6 different subjects. The object of the experiment was to discover the relative effects of speed and intensity of illumination on performance. (*Note.* This is a greatly simplified version of an actual experiment and is given here solely for the purpose of illustrating the analysis of variance with such data.)

TABLE XIII

	Illuminations (A)								
	1			2			3		
	Speeds (B)			Speeds (B)			Speeds (B)		
	1	2	3	1	2	3	1	2	3
Subjects (C) 1	45	38	29	43	35	26	35	29	18
2	38	33	20	40	32	21	34	25	19
3	39	32	21	41	29	25	29	24	16
4	43	37	24	39	30	25	30	27	14
5	40	36	28	42	31	24	31	26	17
6	40	35	25	40	32	22	32	26	16

This table gives a three-way classification as each entry belongs to a certain subject (C) at a certain speed (B) under a certain intensity of illumination (A). The first step in the analysis is to find a convenient working mean and rewrite Table XIII from this mean, inserting totals at the same time. Since the range of readings is from 14 to 45, a convenient working mean is 30. Rewriting the readings about this mean we get Table XIII A.

Table XIII A

	Illuminations (A)												
	1 Speeds (B)				2 Speeds (B)				3 Speeds (B)				
Subjects (C)	1	2	3	Total	1	2	3	Total	1	2	3	Total	Totals
1	15	8	−1	22	13	5	−4	14	5	−1	−12	−8	28
2	8	3	−10	1	10	2	−9	3	4	−5	−11	−12	−8
3	9	2	−9	2	11	−1	−5	5	−1	−6	−14	−21	−14
4	13	7	−6	14	9	0	−5	4	0	−3	−16	−19	−1
5	10	6	−2	14	12	1	−6	7	1	−4	−13	−16	5
6	10	5	−5	10	10	2	−8	4	2	−4	−14	−16	−2
Totals	65	31	−33	63	65	9	−37	37	11	−23	−80	−92	8

The total s.o.s. may be obtained from this table in the usual way, i.e. square each single entry (not the totals), add and subtract $[S(X)]^2/N$. In this case

$$\begin{aligned} \text{total s.o.s.} &= 15^2+8^2+(-1)^2+\ldots \\ &\qquad +2^2+(-4)^2+(-14)^2-8^2/54 \\ &= 3402-1{\cdot}185 \\ &= 3400{\cdot}815. \end{aligned}$$

Main effects

(1) The totals for the A grouping are 63, 37 and −92, each being the sum of 18 readings.

$$\text{Hence}\quad S(\bar{X}_a-\bar{X})^2 = [63^2+37^2+(-92)^2]/18 - 8^2/54$$
$$= 13802/18-1{\cdot}185$$
$$= 765{\cdot}593.$$

(2) The totals for B are obtained by adding together the three totals for B_1, B_2 and B_3 separately, i.e.

$$\begin{aligned} \text{total } B_1 &= 65+65+11 = 141, \\ \text{total } B_2 &= 31+9-23 = 17, \\ \text{total } B_3 &= -33-37-80 = -150. \end{aligned}$$

$$\text{Hence}\quad S(\bar{X}_b-\bar{X})^2 = [141^2+17^2+(-150)^2]/18-8^2/54$$
$$= 42670/18-1{\cdot}185$$
$$= 2369{\cdot}370.$$

(3) The C totals are given at the right of Table XIII A and each contains 9 readings.

Hence

$S(\bar{X}_c - \bar{X})^2$

$= [28^2 + (-8)^2 + (-14)^2 + (-1)^2 + 5^2 + (-2)^2]/9 - 8^2/54$

$= 1074/9 - 1{\cdot}185$

$= 118{\cdot}148.$

First-order interactions

The interactions AB, AC and BC are best found by first extracting three two-way tables from Table XIII A.

(4) That for AB is shown in Table XIII B.

TABLE XIII B

	A_1	A_2	A_3	Totals
B_1	65	65	11	141
B_2	31	9	−23	17
B_3	−33	−37	−80	−150
Totals	63	37	−92	8

Now reference to the previous example shows that, to calculate the s.o.s. for the remainder or interaction in a two-way table, we calculated the total s.o.s. for the table and subtracted from it the sum of the s.o.s. of the main effects. Here we proceed likewise, *bearing in mind that each reading in Table XIII B is the sum of* 6 *readings.* Hence for Table XIII B,

$$\text{the total s.o.s.} = [65^2 + 65^2 + 11^2 + 31^2 + 9^2 + (-23)^2 + (-33)^2 + (-37)^2 + (-80)^2]/6 - 8^2/54$$

$$= 19000/6 - 1{\cdot}185$$

$$= 3165{\cdot}482.$$

We have already calculated the s.o.s. for the main effects A and B, which are involved in this interaction, hence

$$\text{interaction s.o.s. for } AB = 3165{\cdot}482 - 765{\cdot}593 - 2369{\cdot}370 = 30{\cdot}519.$$

(5) Next we extract the two-way table for interaction AC, each reading of which will be the sum of 3 readings. This is shown in Table XIII C.

TABLE XIII C

	A_1	A_2	A_3	Totals
C_1	22	14	−8	28
C_2	1	3	−12	−8
C_3	2	5	−21	−14
C_4	14	4	−19	−1
C_5	14	7	−16	5
C_6	10	4	−16	−2
Totals	63	37	−92	8

From this table, interaction s.o.s. for AC is

$$[22^2+14^2+(-8)^2+\ldots+10^2+4^2+(-16)^2]/3-8^2/54$$
$$-765{\cdot}593-118{\cdot}148$$
$$=2814/3-1{\cdot}185-765{\cdot}593-118{\cdot}148$$
$$=53{\cdot}074.$$

(6) Finally we have, for interaction BC (Table XIII D):

TABLE XIII D

	B_1	B_2	B_3	Totals
C_1	33	12	−17	28
C_2	22	0	−30	−8
C_3	19	−5	−28	−14
C_4	22	4	−27	−1
C_5	23	3	−21	5
C_6	22	3	−27	−2
Totals	141	17	−150	8

This table is obtained by adding the three B_1 entries in Table XIII A for C_1, i.e. $15+13+5=33$, and so on.

From Table XIII D, interaction s.o.s. for BC is

$$[33^3+12^2+(-17)^2+\ldots+22^2+3^2+(-27)^2]/3-8^2/54$$
$$-2369{\cdot}370-118{\cdot}148$$
$$=7506/3-1{\cdot}185-2369{\cdot}370-118{\cdot}148$$
$$=13{\cdot}297.$$

We may now complete our analysis table as under, obtaining the remainder term by adding the s.o.s. for the three main effects and the first-order interactions and subtracting from the total S.O.S. The remainder variance is as usual used as the denominator of the V.R.'s.

S.O.S.		D.F.	Mean squares	V.R.
Between illuminations (A)	765·593	2	382·80	150·7
Between speeds (B)	2369·370	2	1184·68	466·3
Between subjects (C)	118·148	5	23·63	9·3
Interaction AB	30·519	4	7·63	3·0
Interaction AC	53·074	10	5·31	2·09
Interaction BC	13·297	10	1·33	0·52
Remainder	50·814	20	2·54	—
Total	3400·815	53		

11.vi. *Interpretation of analysis.* The most striking feature of this analysis is the very large variance ratios for speeds and illuminations. The data obviously depart significantly from homogeneity—reference to tables is unnecessary with ratios of this size. The experiment shows that both speed and illumination have a very marked effect on performance, especially speed. Differences between subjects are also significant but are of no particular interest. Reference to Fisher and Yates's table shows that the interaction AB is just significant at the 5 % level, but neither of the other first-order interactions is significant. The meaning of this significant interaction AB is that the effects of speed and illumination are not entirely independent.

Before, however, we accept this interaction as important we may apply a further test to it. Since the interactions AC and BC are not significant they may be combined with the 'remainder' to give us a larger number of D.F. for estimating the parent variance. This is done by adding the S.O.S. for the interactions AC and BC and remainder, and also their D.F., which gives us a new remainder variance, estimated with 40 D.F., which may be used as a denominator for the V.R.'s. In general, the more degrees of freedom there are available for estimating a variance, the more reliable is the estimate. The new analysis table produced by this means is given below.

S.O.S.		D.F.	Mean squares	V.R.
Between illuminations (A)	765·593	2	382·80	130·7
Between speeds (B)	2369·370	2	1184·68	404·4
Between subjects (C)	118·148	5	23·63	8·1
Interaction (AB)	30·519	4	7·63	2·60
Remainder	117·185	40	2·93	—
Total	3400·815	53		

Consulting Fisher and Yates's Table V with $N_1 = 4$ and $N_2 = 40$, we find a value for the V.R. of 2·61 for $P = 0{\cdot}05$. This means that now we have employed a better estimate of the parent variance based on a larger number of D.F., the interaction AB is just on the borderline of significance and so cannot be regarded as having an important effect on performance. The final conclusions from the analysis would therefore be that speed has a very marked effect on performance and intensity of illumination also has a marked effect, though not so marked as that of speed, within the range of this particular experiment. The practical value of an experiment such as this is to show that in a human activity requiring reactions of the sort exemplified it is most important to control the speed factor for efficient performance, and also, though in a less degree, to control the factor of illumination. The optimum values of speed and illumination necessary for maximum efficiency cannot be deduced from the above analysis and would have to be the subject of further research.

11.vii. *n*-way classification. Analysis on precisely similar lines to those exemplified above may be made with data which may be classified in 4, 5 or more ways. The amount of arithmetic becomes progressively greater, however. For instance, with a 4-way classification, A, B, C and D, there will be 6 first-order interactions, AB, AC, AD, BC, BD and CD, and 3 second-order interactions, ABC, ABD and BCD, and the remainder will be a *third-order* interaction, $ABCD$. Similarly, with a 5-way classification there will be 10 first-order, 10 second-order and 5 third-order interactions, and the remainder will be a *fourth-order* interaction.

Although the analysis of variance in these cases presents no added difficulties, apart from the greatly increased arithmetical labour, in practice, particularly in the psychological field, the design of an experiment becomes increasingly difficult. If the main effects are to be directly comparable, it is essential that each reading shall be capable of being classified under each separate class heading. This increases the number of tests each subject has to perform and may introduce practice effects and fatigue effects which it may not be possible to isolate.

In the analysis of the previous example the implicit assumption was made that the task required of the subjects was so simple that neither practice nor fatigue would affect performance. This is by no means always the case and very great care has to be exercised in designing psychological experiments where variables such as these may enter. Such devices as taking the *order* in which the tests are done as one of the classes of variates, or taking matched groups of subjects, each of whom does only a fraction of the total tests, are useful on occasion. However, the data in such cases need handling with great care and the elementary student is strongly advised to seek expert help both in designing an experiment so that it will yield data of an appropriate type and also in analysing the results. The normal tests of homogeneity are invalidated if the assumptions, previously mentioned, on which such tests are based are not warranted. It is true that the analysis of variance may be carried out where the data are not normally distributed, where numbers in sub-groups are unequal and even when parts of the data are missing, but homogeneity tests in such cases are complicated and laborious and are in any case beyond the bounds of this simple introduction.

EXERCISE ON CHAPTER XI

38. Re-work and check the arithmetic of Examples **43**, **44** and **45**.

Answers to Exercises

1.

Test	1–25	26–50	51–75	76–100
A	43·20	44·68	44·52	44·28
B	175·76	176·72	193·28	193·72
C	28·76	22·92	23·12	20·64
D	28·64	33·56	29·96	30·00
E	125·00	120·80	124·40	130·32
F	27·48	27·20	26·68	27·72
G	35·08	38·40	36·16	37·52

2. (*a*) 27·34. (*b*) 27·20.

3.

Test	(*a*)	(*b*)	Discrepancy
A	44·14	44·17	0·03
B	185·40	184·87	0·53
C	23·75	23·86	0·08
D	30·38	30·54	0·16
E	125·20	125·13	0·07
F	27·18	27·27	0·09
G	36·75	36·79	0·04

4.

Test	(*a*)	(*b*)	Test	(*a*)	(*b*)
A	45·5	44·0	*E*	123·0	130·0
B	167·0	185·5	*F*	28·0	25·5
C	25·5	22·0	*G*	37·5	39·0
D	30·5	27·5			

5.

Test	(*a*)	(*b*)	Test	(*a*)	(*b*)
A	48·62	43·20	*E*	123·20	135·28
B	148·52	169·50	*F*	29·32	22·10
C	24·82	22·24	*G*	39·02	43·32
D	29·30	22·54			

6.

Test	Lower quartile	Upper quartile	Semi-inter-quartile range
A	38·0	50·0	6·0
B	157·0	203·0	23·0
C	15·0	31·0	8·0
D	19·5	40·0	10·25
E	107·5	142·0	17·25
F	23·0	32·0	4·5
G	29·0	42·0	6·5

7.

Test	(a)	(b)	Test	(a)	(b)
A	7·66	9·76	*E*	17·34	17·32
B	24·08	29·98	*F*	4·74	5·52
C	10·34	8·80	*G*	7·18	6·88
D	9·90	12·50			

8. (*a*) 20·57. (*b*) 15·05.

9.

Test	(a)	(b)	(c)	(d)
A	11·06	8·18	13·77	10·81
C	12·87	11·09	11·93	10·06
D	12·03	12·33	14·78	15·05
F	5·99	6·41	5·90	7·35
G	9·36	8·43	8·17	8·58

10.

Test	S.D.	V.	Test	S.D.	V.
A	11·24	25·46	*E*	21·13	16·88
B	37·47	20·21	*F*	6·46	23·77
C	11·80	49·62	*G*	8·70	23·67
D	13·76	45·29			

11. $\beta_1 = 0{\cdot}000002$; $\beta_2 = 3{\cdot}11$,
$\gamma_1 = 0{\cdot}0014$; S.e. $= 0{\cdot}245$,
$\gamma_2 = 0{\cdot}11$; S.e. $= 0{\cdot}490$.
Hence the distribution does not depart significantly from normality.

12. $\beta_1 = 0{\cdot}0526$; $\beta_2 = 2{\cdot}44$,
$\gamma_1 = 0{\cdot}23$; S.e. $= 0{\cdot}245$,
$\gamma_2 = -0{\cdot}56$; S.e. $= 0{\cdot}490$.
Hence the distribution does not depart significantly from normality.

13. 60.

14. $\frac{1}{2}P = 0{\cdot}0668$. There is therefore no evidence that people were not guessing.

15. Mean difference $= 0{\cdot}12$, $t = 0{\cdot}037$.
Hence the mean difference does not depart significantly from zero.

16.

	Standard error	Difference		Standard error	Difference
(*a*)	1·421	7·39	(*d*)	1·520	3·20
(*b*)	1·813	6·60	(*e*)	1·628	6·37
(*c*)	1·345	3·40	(*f*)	1·084	9·57

All these differences are significant.

17. (*a*) $t = 0{\cdot}52$; difference not significant.
(*b*) $t = 1{\cdot}70$; difference not significant.
(*c*) $t = 1{\cdot}40$; difference not significant.
(*d*) $t = 1{\cdot}29$; difference not significant.
(*e*) $t = 1{\cdot}94$; difference not significant.
(*f*) $t = 4{\cdot}42$; difference significant.
(*g*) $t = 3{\cdot}53$; difference significant.
(*h*) $t = 4{\cdot}25$; difference significant.

18. V.R. = 3·80. This is less than the value of 4·21 approx. given in Fisher and Yates, hence the conclusion is not invalidated.

19. Diff. between proportions = 0·19. S.e. of diff. = 0·0482. Diff. is significant, i.e. influenza is definitely more prevalent in the first town.

20. (*a*) $r = 0{\cdot}215$. (*c*) $r = 0{\cdot}088$.
(*b*) $r = 0{\cdot}246$. (*d*) $r = 0{\cdot}266$.

22.

	A	*B*	*C*	*D*	*E*	*F*	*G*
A	—	0·372	−0·268	−0·262	−0·075	−0·395	−0·158
B	0·372	—	−0·170	−0·068	0·105	−0·057	−0·076
C	−0·268	−0·170	—	0·151	0·236	0·309	0·151
D	−0·262	−0·068	0·151	—	0·032	0·388	0·105
E	−0·075	0·105	0·236	0·032	—	0·338	0·252
F	−0·395	−0·057	0·309	0·388	0·338	—	0·289
G	−0·158	−0·076	0·151	0·105	0·252	0·289	—

This table illustrates a method of tabulating the coefficients of correlation between each pair of a set of tests. For example, the coefficient of correlation between test *C* and test *F* is given in the row headed *C* under the column headed *F*, or in the row headed *F* under the column headed *C*, and is seen to be 0·309.

23. Correlation between *F* and *A*; S.e. = 0·0844,
" " *F* and *B*; S.e. = 0·0997,
" " *F* and *C*; S.e. = 0·0905,
" " *F* and *D*; S.e. = 0·0849,
" " *F* and *E*; S.e. = 0·0886,
" " *F* and *G*; S.e. = 0·0916.

The coefficient differs significantly from zero in every case except that between *F* and *B*.

24. r_{BE} is significantly different from r_{DF} but not from any of the others specified.

25. $r_{DF.C} = 0{\cdot}363$,
$r_{CF.D} = 0{\cdot}275$,
$r_{FG.E} = 0{\cdot}224$,
$r_{EF.G} = 0{\cdot}286$,
$r_{CE.B} = 0{\cdot}259$.

26. (*a*) Between H and C (1–25), $\rho = 0{\cdot}026$,
H and D (1–25), $\rho = 0{\cdot}141$.
(*b*) Between C and D (1–25), $\rho = 0{\cdot}171$,
From Exercise 20 (*a*), $r = 0{\cdot}215$.

27. $\tau = 0{\cdot}440$.

28. Between H and C (1–25), $\tau = -0{\cdot}02$,
H and D (1–25), $\tau = 0{\cdot}128$,
C and D (1–25), $\tau = 0{\cdot}117$.

29. (*a*) For the regression of G on F, $z = 0{\cdot}3387$; hence the regression is not linear. (Alternatively, the variance ratio $= 1{\cdot}97$. That in the Fisher and Yates's 5 % point table for $n_1 = 12$ and $n_2 = 60$ is $1{\cdot}92$; hence the regression departs from linearity.)

For the regression of F on G, the variance due to deviations from linearity is less than that within arrays; hence the regression is linear.

(*b*) $\eta_{yx} = 0{\cdot}530$, $\eta_{xy} = 0{\cdot}394$.

30.

Between	χ^2	C	P	
A and I	6·821	0·253	0·339	
B and I	3·392	0·181	0·757	
C and I	8·987	0·287	0·174	
D and I	13·835	0·349	0·032	Significant
E and I	16·432	0·376	0·012	Significant
F and I	50·085	0·578	0·000001	Significant
G and I	16·352	0·375	0·012	Significant

31.

Between	χ^2	P	
A and I	0·62	0·431	
B and I	0·47	0·493	
C and I	2·52	0·112	
D and I	5·70	0·017	Significant
E and I	10·38	0·0013	Significant
F and I	38·30	0·0000001	Significant
G and I	8·20	0·004	Significant
H and I	5·77	0·016	Significant

32. $\chi^2 = 11{\cdot}58$. $P = 0{\cdot}009$, hence the association is significant.

33. $\chi^2 = 160{\cdot}02$. For this P is exceedingly small. Hence the hypothesis fits the observed data very badly and we conclude that the observer is unreliable. Note that he has an undue tendency to record readings ending in 0 or 5.

34. $y = 3{\cdot}12x + 1{\cdot}585$. Mean square for linear regression is 408·845, that for departures from linear regression is 0·255, hence the fit is exceedingly good.

35. (*a*) $d = 6$. The triads are *ABEA*, *ACEA*, *ADEA*, *BCFB*, *BDFB*, *BEFB*.

(*b*) $\zeta = 0{\cdot}571$.

(*c*) Order is $GA\overline{\underline{BC}}\,\overline{\underline{DE}}F$.

(*d*) The subject is inconsistent in his judgments. This may be because his ability is poor or because the task is too difficult. This could be decided only by further experiments.

36. (*a*) $u = 0{\cdot}556$. $\sqrt{(2\chi^2)} - \sqrt{(2\nu - 1)} = 7{\cdot}05$, hence u is significant.

(*b*) We should conclude that *G*, *A* and *F* are quite distinct in the quality judged, but that there is probably no real difference between *B*, *C*, *D* and *E*.

37. $W = 0{\cdot}853$. $z = 1{\cdot}4284$, hence W is significant.

Table of $\sqrt{\{n_1 . n_2(n_1+n_2-2)/(n_1+n_2)\}}$

	10	11	12	13	14	15	16	17	18	19
10	9·49	9·98	10·44	10·89	11·33	11·75	12·15	12·55	12·93	13·30
11	9·98	10·49	10·98	11·45	11·90	12·34	12·77	13·18	13·58	13·97
12	10·44	10·98	11·49	11·98	12·45	12·91	13·35	13·78	14·20	14·60
13	10·89	11·45	11·98	12·49	12·98	13·46	13·92	14·36	14·80	15·22
14	11·33	11·90	12·45	12·98	13·49	13·98	14·46	14·92	15·37	15·81
15	11·75	12·34	12·91	13·46	13·98	14·49	14·98	15·46	15·93	16·38
16	12·15	12·77	13·35	13·92	14·46	14·98	15·49	15·98	16·46	16·93
17	12·55	13·18	13·78	14·36	14·92	15·46	15·98	16·49	16·99	17·47
18	12·93	13·58	14·20	14·80	15·37	15·93	16·46	16·99	17·49	17·99
19	13·30	13·97	14·60	15·22	15·81	16·38	16·93	17·47	17·99	18·49
20	13·66	14·35	15·00	15·63	16·23	16·82	17·38	17·93	18·47	18·99
21	14·02	14·72	15·39	16·03	16·65	17·25	17·83	18·39	18·94	19·47
22	14·36	15·08	15·76	16·42	17·06	17·67	18·26	18·84	19·40	19·94
23	14·70	15·43	16·13	16·80	17·45	18·08	18·69	19·27	19·84	20·40
24	15·03	15·78	16·49	17·18	17·84	18·48	19·10	19·70	20·28	20·85
25	15·35	16·12	16·85	17·55	18·22	18·87	19·51	20·12	20·71	21·29
26	15·67	16·45	17·19	17·91	18·60	19·26	19·90	20·53	21·14	21·73
27	15·98	16·77	17·53	18·26	18·96	19·64	20·30	20·93	21·55	22·15
28	16·29	17·09	17·87	18·61	19·32	20·01	20·68	21·33	21·96	22·57
29	16·59	17·41	18·19	18·95	19·68	20·38	21·06	21·72	22·36	22·98
30	16·88	17·72	18·52	19·28	20·02	20·74	21·43	22·10	22·75	23·38
31	17·17	18·02	18·83	19·66	20·36	21·09	21·79	22·47	23·13	23·78
32	17·46	18·32	19·15	19·94	20·70	21·44	22·15	22·84	23·52	24·17
33	17·74	18·61	19·45	20·26	21·03	21·78	22·50	23·21	23·89	24·55
34	18·02	18·90	19·76	20·57	21·37	22·12	22·85	23·57	24·26	24·93
35	18·29	19·19	20·05	20·88	21·68	22·45	23·20	23·92	24·62	25·31
36	18·56	19·47	20·35	21·19	22·00	22·78	23·53	24·27	24·98	25·67
37	18·82	19·75	20·64	21·49	22·31	23·10	23·87	24·61	25·33	26·04
38	19·08	20·02	20·92	21·79	22·62	23·42	24·20	24·95	25·68	26·39
39	19·34	20·29	21·20	22·08	22·92	23·73	24·52	25·28	26·03	26·75
40	19·60	20·56	21·48	22·37	23·22	24·05	24·84	25·62	26·37	27·10
41	19·85	20·82	21·76	22·66	23·52	24·35	25·16	25·94	26·70	27·44
42	20·10	21·08	22·03	22·94	23·81	24·66	25·47	26·26	27·03	27·78
43	20·34	21·34	22·30	23·22	24·10	24·96	25·78	26·58	27·36	28·12
44	20·58	21·60	22·56	23·49	24·39	25·25	26·09	26·90	27·68	28·45
45	20·82	21·85	22·83	23·77	24·67	25·54	26·39	27·21	28·01	28·78
46	21·06	22·10	23·08	24·04	24·95	25·83	26·69	27·52	28·32	29·11
47	21·30	22·34	23·34	24·30	25·23	26·12	26·98	27·82	28·63	29·43
48	21·53	22·59	23·60	24·57	25·50	26·40	27·28	28·12	28·94	29·75
49	21·76	22·83	23·85	24·83	25·77	26·68	27·57	28·42	29·25	30·06
50	21·98	23·06	24·10	25·09	26·04	26·96	27·85	28·72	29·56	30·37

APPENDIX A (*cont.*)

	20	21	22	23	24	25	26	27	28	29
10	13·66	14·02	14·36	14·70	15·03	15·35	15·67	15·98	16·29	16·59
11	14·35	14·72	15·08	15·43	15·78	16·12	16·45	16·77	17·09	17·41
12	15·00	15·39	15·76	16·13	16·49	16·85	17·19	17·53	17·87	18·19
13	15·63	16·03	16·42	16·80	17·18	17·55	17·91	18·26	18·61	18·95
14	16·23	16·65	17·06	17·45	17·84	18·22	18·60	18·96	19·32	19·68
15	16·82	17·25	17·67	18·08	18·48	18·87	19·26	19·64	20·01	20·38
16	17·38	17·83	18·26	18·69	19·10	19·51	19·90	20·30	20·68	21·06
17	17·93	18·39	18·84	19·27	19·70	20·12	20·53	20·93	21·33	21·72
18	18·47	18·94	19·40	19·84	20·28	20·71	21·14	21·55	21·96	22·36
19	18·99	19·47	19·94	20·40	20·85	21·29	21·73	22·15	22·57	22·98
20	19·49	19·99	20·47	20·94	21·41	21·86	22·30	22·74	23·17	23·59
21	19·99	20·49	20·99	21·47	21·95	22·41	22·86	23·31	23·75	24·18
22	20·47	20·99	21·49	21·99	22·47	22·95	23·41	23·87	24·32	24·76
23	20·94	21·47	21·99	22·49	22·99	23·47	23·95	24·42	24·87	25·32
24	21·41	21·95	22·47	22·99	23·49	23·99	24·48	24·95	25·42	25·88
25	21·86	22·41	22·95	23·47	23·99	24·49	24·99	25·48	25·95	26·42
26	22·30	22·86	23·41	23·95	24·48	24·99	25·50	25·99	26·48	26·96
27	22·74	23·31	23·87	24·42	24·95	25·48	25·99	26·50	26·99	27·48
28	23·17	23·75	24·32	24·87	25·42	25·95	26·48	26·99	27·50	27·99
29	23·59	24·18	24·76	25·32	25·88	26·42	26·96	27·48	27·99	28·50
30	24·00	24·60	25·19	25·77	26·33	26·88	27·43	27·96	28·48	28·99
31	24·41	25·02	25·62	26·20	26·78	27·34	27·89	28·43	28·96	29·48
32	24·81	25·43	26·04	26·63	27·21	27·78	28·34	28·89	29·43	29·96
33	25·20	25·83	26·45	27·05	27·64	28·22	28·79	29·35	29·89	30·43
34	25·59	26·23	26·86	27·47	28·07	28·66	29·23	29·80	30·35	30·90
35	25·97	26·62	27·26	27·88	28·49	29·08	29·67	30·24	30·80	31·36
36	26·35	27·01	27·65	28·28	28·90	29·50	30·10	30·68	31·25	31·81
37	26·72	27·39	28·04	28·68	29·31	29·92	30·52	31·11	31·69	32·26
38	27·09	27·77	28·43	29·07	29·71	30·33	30·94	31·53	32·12	32·70
39	27·45	28·14	28·81	29·46	30·10	30·73	31·35	31·95	32·55	33·13
40	27·81	28·50	29·18	29·85	30·50	31·13	31·76	32·37	32·97	33·56
41	28·16	28·87	29·55	30·22	30·88	31·53	32·16	32·78	33·39	33·99
42	28·51	29·22	29·92	30·60	31·26	31·92	32·56	33·18	33·80	34·40
43	28·86	29·58	30·28	30·97	31·64	32·30	32·95	33·58	34·21	34·82
44	29·20	29·93	30·64	31·33	32·01	32·68	33·34	33·98	34·61	35·23
45	29·53	30·27	30·99	31·69	32·38	33·06	33·72	34·37	35·01	35·63
46	29·87	30·61	31·34	32·05	32·75	33·43	34·10	34·76	35·40	36·03
47	30·20	30·95	31·69	32·41	33·11	33·80	34·47	35·14	35·79	36·43
48	30·52	31·29	32·03	32·76	33·47	34·16	34·85	35·52	36·17	36·82
49	30·85	31·62	32·37	33·10	33·82	34·52	35·21	35·89	36·56	37·21
50	31·17	31·94	32·70	33·44	34·17	34·88	35·58	36·26	36·93	37·59

APPENDIX A (*cont.*)

	30	31	32	33	34	35	36	37	38	39
10	16·88	17·17	17·46	17·74	18·02	18·29	18·56	18·82	19·08	19·34
11	17·72	18·02	18·32	18·61	18·90	19·19	19·47	19·75	20·02	20·29
12	18·52	18·83	19·15	19·45	19·76	20·05	20·35	20·64	20·92	21·20
13	19·28	19·66	19·94	20·26	20·57	20·88	21·19	21·49	21·79	22·08
14	20·02	20·36	20·70	21·03	21·37	21·68	22·00	22·31	22·62	22·92
15	20·74	21·09	21·44	21·78	22·12	22·45	22·78	23·10	23·42	23·73
16	21·43	21·79	22·15	22·50	22·85	23·20	23·53	23·87	24·20	24·52
17	22·10	22·47	22·84	23·21	23·57	23·92	24·27	24·61	24·95	25·28
18	22·75	23·13	23·52	23·89	24·26	24·62	24·98	25·33	25·68	26·03
19	23·38	23·78	24·17	24·55	24·93	25·31	25·67	26·04	26·39	26·75
20	24·00	24·41	24·81	25·20	25·59	25·97	26·35	26·72	27·09	27·45
21	24·60	25·02	25·43	25·83	26·23	26·62	27·01	27·39	27·77	28·14
22	25·19	25·62	26·04	26·45	26·86	27·26	27·35	28·04	28·43	28·81
23	25·77	26·20	26·63	27·05	27·47	27·88	28·28	28·68	29·07	29·46
24	26·33	26·78	27·21	27·64	28·07	28·49	28·90	29·31	29·71	30·10
25	26·88	27·34	27·78	28·22	28·66	29·08	29·50	29·92	30·33	30·73
26	27·43	27·89	28·34	28·79	29·23	29·67	30·10	30·52	30·94	31·35
27	27·96	28·43	28·89	29·35	29·80	30·24	30·68	31·11	31·53	31·95
28	28·48	28·96	29·43	29·89	30·35	30·80	31·25	31·69	32·12	32·55
29	28·99	29·48	29·96	30·43	30·90	31·36	31·81	32·26	32·70	33·13
30	29·50	29·99	30·48	30·96	31·43	31·90	32·36	32·82	33·26	33·71
31	29·99	30·50	30·99	31·48	31·96	32·44	32·90	33·37	33·82	34·27
32	30·48	30·99	31·50	31·99	32·48	32·96	33·44	33·91	34·37	34·83
33	30·96	31·48	31·99	32·50	32·99	33·48	33·96	34·44	34·91	35·37
34	31·43	31·96	32·48	32·99	33·50	33·99	34·48	34·97	35·44	35·91
35	31·90	32·44	32·96	33·48	33·99	34·50	34·99	35·48	35·97	36·44
36	32·36	32·90	33·44	33·96	34·48	34·99	35·50	35·99	36·48	36·97
37	32·82	33·37	33·91	34·44	34·97	35·48	35·99	36·50	36·99	37·48
38	33·26	33·82	34·37	34·91	35·44	35·97	36·48	36·99	37·50	37·99
39	33·71	34·27	34·83	35·37	35·91	36·44	36·97	37·48	37·99	38·50
40	34·14	34·71	35·28	35·83	36·38	36·91	37·44	37·97	38·48	38·99
41	34·57	35·15	35·72	36·28	36·84	37·38	37·92	38·45	38·97	39·48
42	35·00	35·59	36·16	36·73	37·29	37·84	38·38	38·92	39·45	39·97
43	35·42	36·01	36·60	37·17	37·74	38·29	38·84	39·39	39·92	40·45
44	35·84	36·44	37·03	37·61	38·18	38·74	39·30	39·85	40·39	40·92
45	36·25	36·86	37·45	38·04	38·62	39·19	39·75	40·30	40·85	41·39
46	36·66	37·27	37·87	38·47	39·05	39·63	40·19	40·76	41·31	41·85
47	37·06	37·68	38·29	38·89	39·48	40·06	40·64	41·20	41·76	42·31
48	37·46	38·08	38·70	39·31	39·90	40·49	41·07	41·64	42·21	42·77
49	37·85	38·48	39·11	39·72	40·32	40·92	41·50	42·08	42·65	43·22
50	38·24	83·88	39·51	40·13	40·74	41·34	41·93	42·51	43·09	43·66

APPENDIX A (*cont.*)

	40	41	42	43	44	45	46	47	48	49
10	19·60	19·85	20·10	20·34	20·58	20·82	21·06	21·30	21·53	21·76
11	20·56	20·82	21·08	21·34	21·60	21·85	22·10	22·34	22·59	22·83
12	21·48	21·76	22·03	22·30	22·56	22·83	23·08	23·34	23·60	23·85
13	22·37	22·66	22·94	23·22	23·49	23·77	24·04	24·30	24·57	24·83
14	23·22	23·52	23·81	24·10	24·39	24·67	24·95	25·23	25·50	25·77
15	24·05	24·35	24·66	24·96	25·25	25·54	25·83	26·12	26·40	26·68
16	24·84	25·16	25·47	25·78	26·09	26·39	26·69	26·98	27·28	27·57
17	25·62	25·94	26·26	26·58	26·90	27·21	27·52	27·82	28·12	28·42
18	26·37	26·70	27·03	27·36	27·68	28·01	28·32	28·63	28·94	29·25
19	27·10	27·44	27·78	28·12	28·45	28·78	29·11	29·43	29·75	30·06
20	27·81	28·16	28·51	28·86	29·20	29·53	29·87	30·20	30·52	30·85
21	28·50	28·87	29·22	29·58	29·93	30·27	30·61	30·95	31·29	31·62
22	29·18	29·55	29·92	30·28	30·64	30·99	31·34	31·69	32·03	32·37
23	29·85	30·22	30·60	30·97	31·33	31·69	32·05	32·41	32·76	33·10
24	30·50	30·88	31·26	31·64	32·01	32·38	32·75	33·11	33·47	33·82
25	31·13	31·53	31·92	32·30	32·68	33·06	33·43	33·80	34·16	34·52
26	31·76	32·16	32·56	32·95	33·34	33·72	34·10	34·47	34·85	35·21
27	32·37	32·78	33·18	33·58	33·98	34·37	34·76	35·14	35·52	35·89
28	32·97	33·39	33·80	34·21	34·61	35·01	35·40	35·79	36·17	36·56
29	33·56	33·99	34·40	34·82	35·23	35·63	36·03	36·43	36·82	37·21
30	34·14	34·57	35·00	35·42	35·84	36·25	36·66	37·06	37·46	37·85
31	34·71	35·15	35·59	36·01	36·44	36·86	37·27	37·68	38·08	38·48
32	35·28	35·72	36·16	36·60	37·03	37·45	37·87	38·29	38·70	39·11
33	35·83	36·28	36·73	37·17	37·61	38·04	38·47	38·89	39·31	39·72
34	36·38	36·84	37·29	37·74	38·18	38·62	39·05	39·46	39·90	40·32
35	36·91	37·38	37·84	38·29	38·74	39·19	39·63	40·06	40·49	40·92
36	37·44	37·92	38·38	38·84	39·30	39·75	40·19	40·64	41·07	41·50
37	37·97	38·45	38·92	39·39	39·85	40·30	40·76	41·20	41·64	42·08
38	38·48	38·97	39·45	39·92	40·39	40·85	41·31	41·76	42·21	42·65
39	38·99	39·48	39·97	40·45	40·92	41·39	41·85	42·31	42·77	43·22
40	39·50	39·99	40·48	40·97	41·45	41·92	42·39	42·86	43·32	43·77
41	39·99	40·50	40·99	41·49	41·97	42·45	42·92	43·40	43·86	44·32
42	40·48	40·99	41·50	41·99	42·49	42·97	43·45	43·92	44·40	44·86
43	40·97	41·49	41·99	42·50	42·99	43·49	43·97	44·45	44·93	45·40
44	41·45	41·97	42·49	42·99	43·50	43·99	44·49	44·97	45·45	45·93
45	41·92	42·45	42·97	43·49	43·99	44·50	44·99	45·49	45·97	46·46
46	42·39	42·92	43·45	43·97	44·49	44·99	45·50	45·99	46·49	46·97
47	42·86	43·40	43·92	44·45	44·97	45·49	45·99	46·50	46·99	47·49
48	43·32	43·86	44·40	44·93	45·45	45·97	46·49	46·99	47·50	47·99
49	43·77	44·32	44·86	45·40	45·93	46·46	46·97	47·49	47·99	48·50
50	44·42	44·78	45·32	45·87	46·40	46·93	47·46	47·97	48·49	48·99

Curve showing values of N necessary for different values of r to be significant

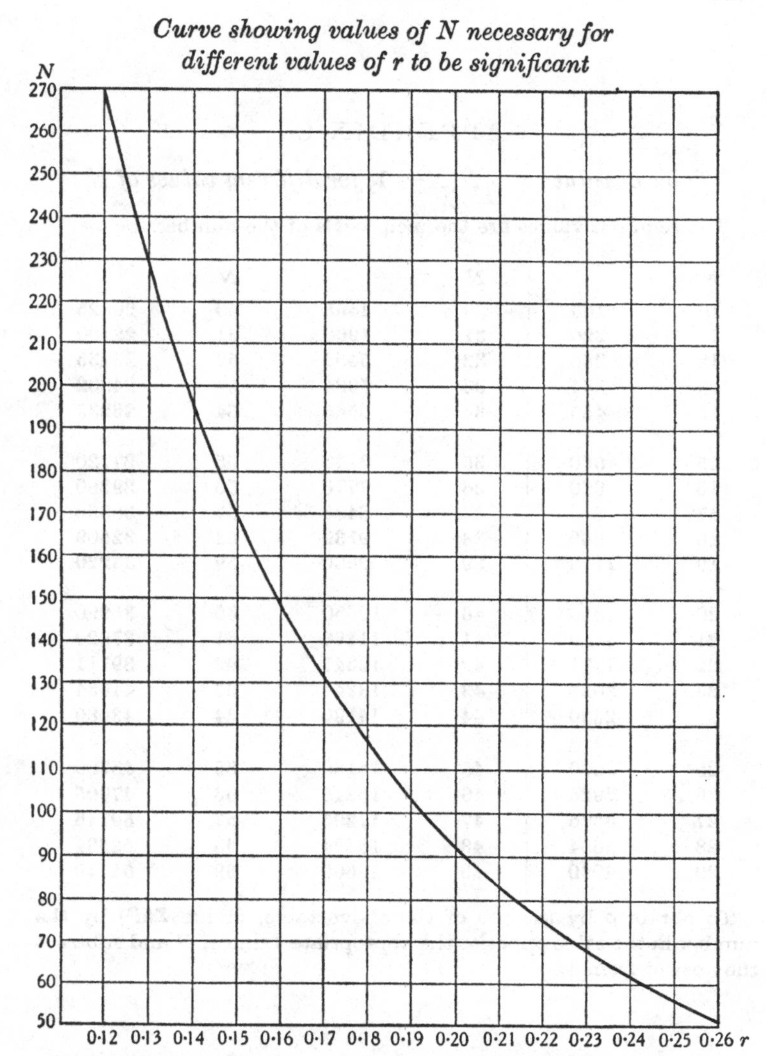

If a calculated r from a sample of N observations is larger than the value given by the curve corresponding to that value of N, it is significant.

Conversely, if the number of observations, N, in a sample yielding a particular value of r is larger than that given by the curve corresponding to that value of r, then r is significant.

In borderline cases the standard error of r should always be calculated as a check.

APPENDIX C

Table of values of $6/N(N^2-1)$ *for different values of* N

The required values are the reciprocals of the numbers below.

N		N		N	
10	165	30	4495	50	20825
11	220	31	4960	51	22100
12	286	32	5456	52	23436
13	364	33	5984	53	24802
14	455	34	6545	54	26235
15	560	35	7140	55	27720
16	680	36	7770	56	29260
17	816	37	8436	57	30856
18	969	38	9139	58	32509
19	1140	39	9880	59	34220
20	1330	40	10660	60	35990
21	1540	41	11480	61	37820
22	1771	42	12341	62	39711
23	2024	43	13244	63	41664
24	2300	44	14190	64	43680
25	2600	45	15180	65	45760
26	2925	46	16215	66	47905
27	3276	47	17296	67	50116
28	3654	48	18424	68	52394
29	4060	49	19600	69	54740

To obtain ρ by the use of the above table, divide $\Sigma(d^2)$ by the number in the table opposite the appropriate value of N and subtract the answer from 1.

APPENDIX D

Squares of numbers from 1 *to* 99

No.	Square	No.	Square	No.	Square	No.	Square
1	1	26	676	51	2601	76	5776
2	4	27	729	52	2704	77	5929
3	9	28	784	53	2809	78	6084
4	16	29	841	54	2916	79	6241
5	25	30	900	55	3025	80	6400
6	36	31	961	56	3136	81	6561
7	49	32	1024	57	3249	82	6724
8	64	33	1089	58	3364	83	6889
9	81	34	1156	59	3481	84	7056
10	100	35	1225	60	3600	85	7225
11	121	36	1296	61	3721	86	7396
12	144	37	1369	62	3844	87	7569
13	169	38	1444	63	3969	88	7744
14	196	39	1521	64	4096	89	7921
15	225	40	1600	65	4225	90	8100
16	256	41	1681	66	4356	91	8281
17	289	42	1764	67	4489	92	8464
18	324	43	1849	68	4624	93	8649
19	361	44	1936	69	4761	94	8836
20	400	45	2025	70	4900	95	9025
21	441	46	2116	71	5041	96	9216
22	484	47	2209	72	5184	97	9409
23	529	48	2304	73	5329	98	9604
24	576	49	2401	74	5476	99	9801
25	625	50	2500	75	5625		

APPENDIX E

Numerical material for exercises

The following two pages of figures give the scores of 100 subjects in certain psychological tests. The scores in seven such tests are given in columns A to G inclusive. Column H gives the ranked order of merit of the subjects in a scholastic examination, and column I gives an assessment of the practical ability of the subjects in some handicraft. This assessment is in four categories—Very Good (V.G.), Good (G.), Fair (F.) and Poor (P.).

These two pages of scores provide the basis for most of the exercises at the end of the successive chapters of the text. The student who wishes for additional exercises may easily make up others for himself from the figures given.

Subject	*A*	*B*	*C*	*D*	*E*	*F*	*G*	*H*	*I*
1	41	164	31	45	121	31	29	50	V.G.
2	49	168	13	30	145	31	35	12	G.
3	41	132	20	5	108	25	25	73½	F.
4	35	165	31	25	141	30	40	28½	G.
5	35	152	30	27	96	18	28	92	P.
6	55	156	45	33	154	35	49	4	V.G.
7	50	153	38	41	102	27	40	94	F.
8	38	165	42	62	156	31	26	11	G.
9	49	161	30	24	147	28	39	30	F.
10	46	156	30	37	124	27	58	52	F.
11	28	167	30	23	98	19	33	93	F.
12	64	204	46	15	92	19	18	96	P.
13	48	203	36	35	170	36	41	1	G.
14	31	139	2	38	126	29	35	52	G.
15	45	191	41	19	127	28	38	52	F.
16	49	156	44	11	131	22	33	35	F.
17	66	240	18	20	111	17	36	60	P.
18	46	146	26	40	105	26	21	75	P.
19	21	138	44	31	122	39	27	54	V.G.
20	27	180	30	30	100	33	50	84	G.
21	50	277	19	27	152	34	39	18	G.
22	49	174	5	16	125	25	27	55	F.
23	26	176	44	38	116	34	40	73½	G.
24	39	157	15	29	125	19	45	56	G.
25	52	274	9	15	131	24	25	36	P.
26	48	167	13	10	94	22	47	95	F.
27	44	152	7	42	80	30	30	100	V.G.
28	50	187	25	36	109	24	41	87	F.
29	49	193	12	32	114	31	23	70	P.
30	45	196	30	40	158	26	38	7½	G.
31	61	184	6	10	94	25	26	97	F.
32	49	222	22	28	137	27	40	31	G.
33	38	196	24	53	127	36	34	57½	V.G.
34	41	188	31	35	137	30	37	37	G.
35	50	234	15	28	107	36	45	72	V.G.
36	52	201	6	43	121	30	34	57½	F.
37	32	143	39	37	142	32	51	33	G.
38	35	177	32	35	141	28	40	33	F.
39	44	156	21	30	115	19	39	76	P.
40	61	245	20	49	121	29	42	59	V.G.
41	39	153	42	43	106	27	27	77	F.
42	52	181	31	43	109	29	45	71	V.G.
43	47	150	19	20	129	27	55	38½	G.
44	29	146	14	59	133	35	42	38½	V.G.
45	48	129	23	25	107	8	23	82	P.
46	40	173	28	18	149	25	54	13	P.
47	44	164	17	30	158	13	40	5	P.
48	29	164	53	43	90	33	37	85	G.
49	50	177	22	15	141	32	34	33	V.G.
50	40	140	21	35	101	26	36	86	F.

Subject	*A*	*B*	*C*	*D*	*E*	*F*	*G*	*H*	*I*
51	72	262	4	54	116	17	22	61	P.
52	71	160	22	17	128	28	44	40	V.G.
53	51	247	4	10	130	23	40	41	F.
54	38	205	45	34	133	29	51	43	V.G.
55	58	266	22	26	126	19	26	42	P.
56	49	210	34	25	141	23	39	19	G.
57	59	139	5	27	123	24	39	49	F.
58	31	162	27	38	142	24	35	21	P.
59	40	178	17	7	84	24	33	91	P.
60	59	149	26	25	118	24	20	63	G.
61	34	150	27	28	90	19	27	99	F.
62	25	216	36	34	157	39	37	9	V.G.
63	39	195	24	12	151	22	29	7½	P.
64	51	216	25	57	99	25	25	88	F.
65	50	178	27	21	146	31	30	21	G.
66	35	283	43	31	162	35	44	2	F.
67	15	170	19	50	127	36	39	63	V.G.
68	40	157	39	14	139	21	45	16½	P.
69	24	144	12	43	82	26	43	98	F.
70	57	236	5	16	135	21	33	21	G.
71	48	185	19	13	124	32	39	63	F.
72	34	146	21	56	148	37	49	15	V.G.
73	45	169	35	35	96	29	40	89	V.G.
74	37	182	7	25	90	26	31	90	F.
75	51	227	33	51	123	33	44	65	V.G.
76	23	149	11	46	130	32	46	44	V.G.
77	45	244	14	34	113	25	39	66	F.
78	51	228	20	15	143	32	42	23	G.
79	41	172	48	60	127	37	40	45	G.
80	63	216	20	9	137	16	26	24	P.
81	39	170	7	42	107	27	39	78	G.
82	59	234	5	22	96	25	32	79	F.
83	45	206	31	37	147	31	40	25	G.
84	67	180	12	9	100	12	44	82	P.
85	42	170	22	32	137	19	55	16½	F.
86	39	198	12	20	162	23	35	3	F.
87	34	200	17	14	145	27	28	26	G.
88	39	148	32	30	136	35	27	46	G.
89	20	163	33	35	153	33	28	10	F.
90	43	187	33	58	152	41	53	14	V.G.
91	37	174	29	17	142	40	47	27	G.
92	42	322	21	40	102	23	37	68	P.
93	46	186	24	55	145	35	41	28½	G.
94	54	214	21	19	130	30	27	47	F.
95	40	167	27	15	116	25	45	67	P.
96	45	159	23	25	142	37	44	80	V.G.
97	52	203	7	34	96	21	21	82	F.
98	35	197	18	20	123	22	40	48	F.
99	47	167	7	48	115	24	29	69	G.
100	59	189	22	14	162	21	33	6	F.

APPENDIX F

Values of $n(n-1)(2n+5)$ for different values of n

n	$n(n-1)(2n+5)$	n	$n(n-1)(2n+5)$	n	$n(n-1)(2n+5)$
2	18	22	22638	42	153258
3	66	23	25806	43	164346
4	156	24	29256	44	175956
5	300	25	33000	45	188100
6	510	26	37050	46	200790
7	798	27	41418	47	214038
8	1176	28	46116	48	227856
9	1656	29	51156	49	242256
10	2250	30	56550	50	257250
11	2970	31	62310	51	272850
12	3828	32	68448	52	289068
13	4836	33	74976	53	305916
14	6006	34	81906	54	323406
15	7350	35	89250	55	341550
16	8880	36	97020	56	360360
17	10608	37	105228	57	379848
18	12546	38	113886	58	400026
19	14706	39	123006	59	420906
20	17100	40	132600	60	442500
21	19740	41	142680		

This table may be used in calculating the significance of τ.

APPENDIX G

Values of $t(t-1)(t-2)$ for different values of t

t	$t(t-1)(t-2)$	t	$t(t-1)(t-2)$	t	$t(t-1)(t-2)$
3	6	19	5414	35	39270
4	24	20	6840	36	42840
5	60	21	7980	37	46620
6	120	22	9240	38	50616
7	210	23	10626	39	54834
8	336	24	12144	40	59280
9	504	25	13800	41	63960
10	720	26	15600	42	68880
11	990	27	17550	43	74046
12	1320	28	19656	44	79464
13	1716	29	21924	45	85140
14	2184	30	24360	46	91080
15	2730	31	26970	47	97290
16	3360	32	29760	48	103776
17	4080	33	32736	49	110544
18	4896	34	35904	50	117600

This table may be used for calculating the significance of τ when there are tied rankings.

APPENDIX H

Values of $mn(m-1)(n-1)/4$ for different values of n and m

n	$m = 2$	$m = 3$	$m = 4$	$m = 5$	$m = 6$
3	3	9	18	30	45
4	6	18	36	60	90
5	10	30	60	100	150
6	15	45	90	150	225
7	21	63	126	210	315
8	28	84	168	280	420
9	36	108	216	360	540
10	45	135	270	450	675
11	55	165	330	550	825
12	66	198	396	660	990
13	78	234	468	780	1170
14	91	273	546	910	1365
15	105	315	630	1050	1575

To calculate u, divide 2Σ by the appropriate number above and subtract 1.

APPENDIX I

Values of ν for different values of n and m

n	$m = 3$	$m = 4$	$m = 5$	$m = 6$
3	18	9	$6{\cdot}\dot{6}$	5·625
4	36	18	$13{\cdot}\dot{3}$	11·25
5	60	30	$22{\cdot}\dot{2}$	18·75
6	90	45	$33{\cdot}\dot{3}$	28·125
7	126	63	$46{\cdot}\dot{6}$	39·375
8	168	84	$62{\cdot}\dot{2}$	52·5
9	216	108	80	67·5
10	270	135	100	84·375
11	330	165	$122{\cdot}\dot{2}$	103·125
12	396	198	$146{\cdot}\dot{6}$	123·75
13	468	234	$173{\cdot}\dot{3}$	146·25
14	546	273	$202{\cdot}\dot{2}$	170·625
15	630	315	$233{\cdot}\dot{3}$	196·875

This table may be used in the testing of the significance of u.

APPENDIX J

Values of $\frac{1}{2}\binom{n}{2}\binom{m}{2}\frac{m-3^*}{m-2}$ *for different values of n and m*

n	$m=4$	$m=5$	$m=6$
2	1·5	3·3̇	5·625
3	4·5	10	16·875
4	9	20	33·75
5	15	33·3̇	56·25
6	22·5	50	84·375
7	31·5	70	118·125
8	42	93·3̇	157·5
9	54	120	202·5
10	67·5	150	253·125
11	82·5	183·3̇	309·375
12	99	220	371·25
13	117	260	438·75
14	136·5	303·3̇	511·875
15	157·5	350	590·625

This table may be used in calculating the significance of u.

* For $m = 3$ this expression equals 0 for all values of n.

APPENDIX K

Values of $m^2(n^3-n)/12$ *for different values of n and m*

n	$m=3$	$m=4$	$m=5$	$m=6$
3	18	32	50	72
4	45	80	125	180
5	90	160	250	360
6	157·5	280	437·5	630
7	252	448	700	1008
8	378	672	1050	1512
9	540	960	1500	2160
10	742·5	1320	2062·5	2970
11	990	1760	2750	3960
12	1287	2288	3575	5148
13	1638	2912	4550	6552
14	2047·5	3640	5687·5	8190
15	2520	4480	7000	10080

To find W, divide S by the appropriate number in the above table. If there are ties, divide S by the appropriate number from the table *less* $m\Sigma(T)$.

Index

References are to pages